Advances in Industrial Control

Advances in Industrial Control is a series of monographs and contributed titles focusing on the applications of advanced and novel control methods within applied settings. This series has worldwide distribution to engineers, researchers and libraries.

The series promotes the exchange of information between academia and industry, to which end the books all demonstrate some theoretical aspect of an advanced or new control method and show how it can be applied either in a pilot plant or in some real industrial situation. The books are distinguished by the combination of the type of theory used and the type of application exemplified. Note that "industrial" here has a very broad interpretation; it applies not merely to the processes employed in industrial plants but to systems such as avionics and automotive brakes and drivetrain. This series complements the theoretical and more mathematical approach of Communications and Control Engineering.

Indexed by SCOPUS and Engineering Index.

Series Editors

Professor **Michael J. Grimble**
Department of Electronic and Electrical Engineering, Royal College Building, 204 George Street, Glasgow G1 1XW, United Kingdom
e-mail: m.j.grimble@strath.ac.uk

Professor **Antonella Ferrara**
Department of Electrical, Computer and Biomedical Engineering, University of Pavia, Via Ferrata 1, 27100 Pavia, Italy
e-mail: antonella.ferrara@unipv.it

or the

In-house Editor

Mr. **Oliver Jackson**
Springer London, 4 Crinan Street, London, N1 9XW, United Kingdom
e-mail: oliver.jackson@springer.com

Publishing Ethics

Researchers should conduct their research from research proposal to publication in line with best practices and codes of conduct of relevant professional bodies and/or national and international regulatory bodies. For more details on individual ethics matters please see:
https://www.springer.com/gp/authors-editors/journal-author/journal-author-helpdesk/publishing-ethics/14214

More information about this series at http://www.springer.com/series/1412

Péter Gáspár · Balázs Németh

Predictive Cruise Control for Road Vehicles Using Road and Traffic Information

Springer

Péter Gáspár
MTA SZTAKI
Budapest, Hungary

Balázs Németh
MTA SZTAKI
Budapest, Hungary

ISSN 1430-9491 ISSN 2193-1577 (electronic)
Advances in Industrial Control
ISBN 978-3-030-04115-1 ISBN 978-3-030-04116-8 (eBook)
https://doi.org/10.1007/978-3-030-04116-8

Library of Congress Control Number: 2018960760

This Springer imprint is published by the registered company Springer Nature Switzerland AG
The registered company address is: Gewerbestrasse 11, 6330 Cham, Switzerland

Series Editor's Foreword

Control systems engineering is viewed very differently by researchers and those that practice the craft. The former group develops general algorithms with a strong underlying mathematical basis while for the latter, concerns over the limits of equipment and plant downtime dominate. The series *Advances in Industrial Control* attempts to bridge this divide and hopes to encourage the adoption of more advanced control techniques when warranted.

The rapid development of new control theory and technology has an impact on all areas of control engineering and applications. There are new control theories, actuators, sensor systems, computing methods, design philosophies, and of course new application areas. This provides justification for a specialized monograph series, and the development of relevant control theory also needs to be stimulated and driven by the needs and challenges of applications. A focus on applications is also essential if the different aspects of the control design problem are to be explored with the same dedication the synthesis problems have received. The series provides an opportunity for researchers to present an extended exposition of new work on industrial control, raising awareness of the substantial benefits that can accrue, and the challenges that can arise.

The authors are well known for their work on vehicle control systems, driver assistance systems, and traffic flow. This book is concerned with the design of an automated longitudinal control system for vehicles to enhance the capabilities of adaptive cruise control systems. There are two optimization problems where a balance in performance is required involving the longitudinal control force to be minimized and the traveling time that must also be minimized. There is clearly a conflict in the wish to minimize energy whilst reducing journey times so a natural optimization problem arises. It is assumed that the vehicle has information about the environment and surrounding vehicles which is much easier to achieve with recent developments in sensor technology for autonomous vehicles. The predictive cruise control aims to balance the need for energy saving against journey time according to the needs of the driver. The major sections of the text cover *Predictive Cruise Control*, the *Analysis of the Traffic Flow*, and *Control Strategies*.

There is a huge interest in all aspects of vehicle control systems and traffic flow control. This book covers many of the important topics such as traffic and platoon control, and it describes the main areas of control methodologies, modeling, design, simulation, and results. The main focus of the book is to ensure that the velocity of the vehicle is controlled so that the global and local information about traveling and the environment is taken into consideration. Such work is clearly important for both safety and the environment, and it is therefore a welcome addition to the series on *Advances in Industrial Control*.

Glasgow, UK
October 2018

Michael J. Grimble

Contents

Chapter 1
Introduction

Introductory Thoughts

The automation of road transport systems has recently become the main focus of researchers and automotive companies as well. Several car manufacturers have already introduced autonomous vehicle functions which can be regarded as milestones in the development of fully autonomous or self-driving vehicles. Research on next generation of adaptive cruise control and cooperative adaptive cruise control systems generally focuses on enhancing the performances of the system by considering driver behavior.

In particular, the development of an energy-efficient operation strategy for road vehicles has been in the focus. The purpose of the strategy is to design the speed of road vehicles taking into consideration several factors such as control energy requirement, fuel consumption, road slopes, speed limits, emissions, and traveling time. These optimization criteria lead to multi-objective solutions.

Certainly, other approaches are also used. In crossing an intersection, the most important consideration is to ensure the continuity of the traffic, i.e., the continuity of the passage of cars. If a car needs to slow down or stop at an intersection due to other traffic, the capacity of the road decreases, the average speed of vehicles decreases, and fuel consumption increases. If the continuity of traffic can be guaranteed by using appropriately tuned traffic lights or other solutions, the abovementioned factors for the speed optimization are applied again.

The book focuses on the design of a multi-criteria automated vehicle longitudinal control system as an enhancement of the adaptive cruise control system. As in most of the longitudinal automated vehicle control systems, it is assumed that the vehicle has information about the environment and surrounding vehicles using wireless or Cloud-Based Vehicle-to-Infrastructure and Vehicle-to-Vehicle (V2I and V2V) communication technologies. In the speed design both the road and the traffic information is also taken into consideration. This leads to the predictive cruise

P. Gáspár and B. Németh, *Predictive Cruise Control for Road Vehicles Using Road and Traffic Information*, Advances in Industrial Control,
https://doi.org/10.1007/978-3-030-04116-8_1

control, which is able to create a balance between longitudinal energy saving and journey time according to preferences of the driver.

However, other drivers on the road have different priorities, which can lead to conflict, e.g., fast vehicles are held up by vehicles traveling in a fuel efficient fashion. The difficulty in the predictive speed design is to adapt to the motion of the surrounding vehicles. Making a decision to change lanes is a critical one, in which the conflicts between vehicles and tailbacks must be eliminated. Handling the preceding vehicle and considering the motion of the follower vehicle must be incorporated into the decision method. The combination of the concept of the predictive speed and the congestion problem leads to a more complex multi-criteria optimization task.

There is a strong interaction between the traffic flow and the individual vehicles. This interaction is analyzed from both a microscopic and a macroscopic point of view. According to the microscopic view, the vehicle equipped with predictive control has impact on the traffic flow, which differs from the human-driven vehicles. The parameter variations of the predictive control are analyzed through a sensitivity analysis.

In the macroscopic view, the individual vehicle is incorporated into the global traffic flow. The control of the macroscopic traffic flow and that of the individual microscopic vehicles are handled simultaneously. The purpose is to analyze the effects of different parameters on the average traffic speed and the traction force of the vehicles in the mixed traffic flow by using a macroscopic point of view. The control of the individual vehicles and the traffic control are handled simultaneously, consequently, a trade-off between the parameters of the microscopic and the macroscopic models has been achieved. The purposes of the control design are to avoid congestion through the stability of the system, minimize energy consumption, and reduce the queue length at the control gates.

Another important analysis is related to the platoon control, in which a group of vehicles are traveling at the same speed together. This speed is realized by the leader vehicle, which is followed by the other vehicles. Consequently, the common speed usually deviates from the optimal speed of the individual vehicles. The main task in the design phase is to determine the common speed at which the velocities of the members are as close as possible to their own optimal velocity. Here, the stability analysis of the platoon control in which the predictive control design is used in the individual vehicles is a critical task.

The speed control proposed in the book is analyzed and verified both in a simulation environment and in real circumstances. These solutions and their results will also be presented in the book.

1.1 Motivation Background Concerning Autonomous Vehicle Control

The main motivation of the research and development was the autonomous (or self-driving) cars. Nowadays, the automotive industry is changing continuously, affecting nearly almost every area of development. Concerning the powertrain system, alternative solutions such as hybrid and electric drives are spreading slowly but steadily. This process was further accelerated by the "diesel scandal", which exploded in 2015 and by the verdict of the German federal court in February 2018, which allowed the ban of diesel vehicles with an environmental category lower than Euro 6.

A fast developing area is the Advanced Driver Assistance Systems (ADAS). The original purposes of ADAS systems were to design and implement components and functions to support the driver in the driving process and enhance safety, see, e.g., Gáspár et al. (2017), Sename et al. (2013). The goals of the researchers and developers today are to increase the levels of automated solutions and prepare functions and components to achieve fully automated vehicles to travel on roads. These developments have had a great impact on two technology areas. One is modern wireless info-communication solutions, and the other is artificial intelligence, including machine learning. Since traditional car makers and suppliers have had no prior knowledge of these areas, large IT companies are presented with great possibilities in the vehicle industry. In recent years, this has had a profound effect on developments, among which there are positive and negative examples.

One of the most significant developments is Google's self-driving cars, which are tested in certain cities in Arizona as part of a public pilot project called Waymo, see Waymo (2017). Developers are very serious about safety and both virtual and real-world tests.

Unfortunately, negative experiences have also been found in recent years. One, which is related to the Tesla Autopilot system, has led to a fatal accident. In another sad incident, Uber's self-test vehicle under human supervision run over a bicycle. Because of the hot topic of the autonomous vehicles, developers try to produce results as quickly as possible and do not always follow the security and testing procedures that have been proven by traditional vendors. All of these raise serious ethical issues that could jeopardize the social acceptance of self-driving vehicles.

An autonomous car (also known as a self-driving car) is a vehicle that is capable of sensing its environment, evaluating the real situation, making decision without human interventions, and moreover, activating the components of actuators. Regarding autonomous vehicles, three main tasks to be solved must be highlighted. The first is sensing the environmental, in which a space around the vehicle is monitored continuously applying several sensors and sensor fusion methods. Its purpose is to achieve the most accurate and reliable model of the environment. The second is the situation assessment, in which the system evaluates the given traffic situation based on the environmental situation in order to prepare an adequate decision. This is usually complemented by making the appropriate decision on the maneuvre required in the given situation. The third task is to design a vehicle control and implement it in a safe and reliable way.

In order to determine to what extent the components and functions of different manufacturers and suppliers are suitable for self-drive vehicles, Society of Automotive Engineers (SAE) has introduced a six-level system of requirements in Recommendation J3016. Although the currently implemented autonomous components are at level 2, manufacturers and suppliers are promising levels 4 and 5 within 5–10 years. Moreover, it is important to note that the current transport environment is designed for human perception. The human abilities and experience that a driver uses are extremely difficult to create by using different software systems. There are many unclear or even contradictory traffic situations on the roads. These tasks are often solved by the drivers in an intuitive way and/or by having interaction with the other participants in the traffic. Special situations are very difficult to handle in an automated manner, therefore much clearer traffic rules and better controlled infrastructure are required.

In the current trends, the topics of electromobility and autonomous vehicles have priority. In the former, a partially solved and relatively well-defined problem, i.e., energy storage, should be managed. In the latter topic, there are a large number of unsolved problems concerning regulatory and ethical issues. Nevertheless, in both areas, manufacturers have ambitious plans for a similar span of time, claiming that within a year, level 3 functions and systems will appear, and between 2020 and 2025, levels 4 and 5. However, level 3 systems are still not available in mass production. Accordingly, prediction and promises concerning level 4 and especially the level 5 are welcome with serious reservations. As an example in the Waymo project, a set of cars can be used by volunteer drivers. These vehicles only travel within the cities but completely autonomously without human intervention. This is foreseeing that within a few years, even though a limited area, autonomous vehicles, which can be used by anyone, will appear.

As far as the research and development directions are concerned, the picture is much clearer. In the field of sensors, it is typical that all manufacturers want to cover their vehicles in full space (360°) in a redundant manner, multirange and viewing angles. Manufacturers require technologies in which camera, radar, ultrasound, and lidar sensors are applied simultaneously. Some developers are trying to handle tasks using a pure camera-based solution but they must prove the acceptable reliability. The first three technologies have already become widespread in vehicles owing to their low cost. Although the price of lidar sensors is steadily decreasing, it is still too expensive for mass production. The sensor sets differ with each manufacturer, but there is a broad consensus in the principles. Another important trend where developers' views are also relatively consensual is the application of artificial intelligence, e.g., machine learning methods, in the new complex tasks. Almost everyone agrees that the current rule-based algorithms alone cannot solve all complex perception, situational assessment, and control tasks.

In the forefront of research and development are autonomous functions. The challenge is that the control systems of vehicles must be synchronized with the environment. In the task focused on in the book, the velocity of the vehicle must be designed and implemented in such a way that global and local information about traveling and the environment is taken into consideration. Global information may include the required driving/delivery time, fuel consumption, road slopes, road conditions, speed limits, road stability, and safety. Local information is the speeds of the vehicles on the road, congestions, but also road constructions affecting speed. As a vehicle with speed control is a participant in traffic, it is likely to affect the traveling of vehicles in its environment, but these vehicles also influence the speed design. It is important that the vehicle with speed control must not interfere with or threaten the continuous and safe travel other vehicles involved in the traffic. The vehicle has different impacts during traveling that must be taken into account when driving, e.g., the slower speed of the vehicle ahead of it, the higher speed of the vehicle behind it, and the congestion of the traffic.

The introduction of new technologies poses challenges to be met. The successful algorithms must be tested and validated, which will be a huge task for developers and approval authorities. During the testing, situation-based cases must be examined instead of functional cases. According to the industry's estimation, it requires several million of kilometers of testing, which is time consuming and expensive. Moreover, this requirement encourages the simultaneous application of simulation-based solutions. Another new problem to be solved is the safety of the Wireless Technology (Connected Car) used by autonomous cars. These systems are currently found in the entertainment and comfort features of vehicles, which can be used to connect personal "smart devices" to the vehicle. An important area of applications is Vehicle-to-Vehicle (V2V) and Vehicle-to-Infrastructure (V2I) networks. Using these networks, vehicles are able to exchange their driving dynamics and remotely access the infrastructure signals and status. In this way, vehicles are able to increase the reliability of sensor data and even give new tools to the authorities in traffic control or enforcement. At the same time, it must be accepted as a fact that wireless communication is physically "open". Consequently, the protection of property and personal data will be a new safety task. Moreover, it is necessary to prepare for attacks that can cause traffic anomalies or even accidents.

1.2 Structure of the Book

The book is organized as follows. Chapter 1 presents the motivation background of the research and development of the speed control.

The book is organized around three main parts. The first part focuses on the basis of the predictive cruise control, see Part I.

Chapter 2 presents the basics of the predictive cruise control. The purposes of the speed design are to reduce longitudinal energy requirement and fuel consumption while traveling time remains as short as possible. In the calculation, the road slopes, the speed limits, and the average speeds of the road sections are taken into consideration. By choosing the appropriate velocity according to the road and traffic information, the number of unnecessary accelerations and brakings, moreover, their durations can be significantly reduced. The cruise control design leads to two optimization problems: the longitudinal control force must be minimized; the traveling time must be minimized. In the design, a balance between the two performances must be achieved.

In the design of predictive cruise control, road and traffic information must be taken into consideration. This is the subject of Chap. 3. However, other drivers on the road have different priorities, which can lead to conflict. For example, since the vehicle may catch up with a preceding vehicle, it is necessary to consider its speed. In another example, since the vehicle preferring energy saving is traveling in traffic, it may be in conflict with other vehicles preferring cruising at the speed limit. The goal of the research is to design an optimal predictive control strategy which is able to adapt to the motion of the surrounding vehicles. The combination of the predictive cruise control concept and the congestion problem leads to a complex multi-criteria optimization task. Moreover, a decision algorithm of the lane change is developed. During the lane change, safe operation must be guaranteed and the conflicts between vehicles and tailbacks must be prevented. Handling the preceding vehicle and considering the motion of the follower vehicle must be incorporated into the decision method.

Chapter 4 focuses on the conflict situations in intersections, in which both the safety and the energy-efficient motion of the traffic must be simultaneously guaranteed. However, if a fault occurs in an infrastructure element, these criteria cannot be guaranteed by the traffic control system. The method uses an energy-optimal look-ahead algorithm which considers the motion of the other vehicles, topographic, and road information. The operation of the vehicle control results in an energy-efficient cruising of the controlled vehicle, adapting to the priorities of the other vehicles in the intersection.

The second part focuses on the analysis of the traffic flow both in microscopic and macroscopic point of view, see Part II.

Chapter 5 analyzes the relationship between the traffic flow and the cruise control from the microscopic point of view. There is a relationship between the traffic flow and the predictive cruise control, i.e., they interact strongly with each other. Since the speeds of the individual vehicles affect the speed of the traffic flow, a sensitivity analysis of the parameter variation in the predictive control is performed. If traffic information is also considered in the predictive control, an undesirable side effect on the traffic flow may occur. Therefore, in the cruise control design, both the individual energy optimization and its impact on the traffic flow are elaborated. A method is developed by which the unfavorable effect of the traffic flow consideration can be reduced. In the simulation examples, the speed design is performed in Matlab/Simulink while the analysis is carried out in CarSim and TruckSim simulation and visualization environments.

Chapter 6 analyzes the impact of cruise control on the traffic flow from the macroscopic point of view. The model of the macroscopic traffic flow, the control of traffic dynamics, and the optimization of the individual microscopic vehicles are coordinated. Thus, the individual vehicle is incorporated into the global traffic flow. Since the speed profile of the vehicle equipped with predictive speed control may differ from that of the conventional vehicle, the characteristics of the traffic flow change. The purpose is to analyze the effects of different parameters on the average traffic speed and the traction force of the vehicles in the mixed traffic flow by using a macroscopic point of view. Three components of the traffic system are chosen, such as the inflow of the vehicles on the highway section, the ratio of the vehicles equipped with speed control in the entire traffic, and the energy-efficient parameter of the design of the predictive cruise control. In the analysis, the VisSim simulation environment is applied.

The third main part develops several control strategies for the ramp metering control of the traffic dynamics and presents briefly the implementation of the speed control, see Part III.

The macroscopic modeling and dynamic analysis of the mixed traffic flow, the ramp metering control of the traffic dynamics, and the optimization of the predictive cruise control of the microscopic individual vehicles are coordinated in Chap. 7. The control of the individual vehicles and the traffic control are handled simultaneously, consequently, a trade-off between the parameters of the microscopic and the macroscopic models has been achieved. The purposes of the stability control are to avoid congestion, minimize energy consumption, and reduce the queue length at the control gates. The so-called maximum controlled invariant set provides a stability analysis of the traffic system and calculates the maximum vehicle number which can enter the traffic network. This control system guarantees both the stability of the entire traffic and energy and time optimal intervention of automated vehicles.

In Chap. 8, a design method is developed in which the control of the macroscopic traffic flow and the cruise control of the local vehicles are coordinated. The contribution will be an optimization strategy, which incorporates the nonlinearities and the parameter dependency of the traffic system and the multi-optimization of the look-ahead vehicles. Consequently, a trade-off between the parameters of the microscopic and the macroscopic models has been created. In the method, the impact of traffic and vehicle parameters on the fundamental diagram is analyzed. In the control design, the MPC method is applied, with which the prediction of the traffic flow and that of the traveling of the vehicles are taken into consideration.

Chapter 9 focuses on data-driven coordination design of traffic control. The motivation is that the control of the traffic flow based on the classical state space representation for mixed traffic can be difficult due to the uncertainties, which leads to a data-driven approach. A data-driven coordinated traffic and vehicle control strategy is proposed, with which the inflow at the entrance gates and the speed profile of the eco-cruise controlled vehicles are influenced. Thus, the intervention possibilities are the green time of the traffic lights on the entrances and the speed profile of the cruise controlled vehicles. The advantage of the method is that in the proposed strategy, the fundamental diagram of the traffic dynamics, which contains several parameter uncertainties, is avoided.

In Chap. 10, the method is extended to vehicles in a platoon. The main idea behind the design is that each vehicle in the platoon is able to calculate its speed independently of the other vehicles. Since traveling in a platoon requires the same reference speed, the optimal speed must be modified according to the other vehicles.

In the platoon, the speed of the leader vehicle determines the speed of all the vehicles. The goal is to determine the common speed at which the cruising of the members is as close as possible to their respective optimal speed. The stability analysis of the platoon control in which the predictive cruise control is designed by using the speed control of the individual vehicles must be performed.

Chapter 11 focuses on the simulation and validation of the predictive cruise control. In order to analyze the operation of the predictive cruise control, a Hardware-in-the-Loop vehicle simulator has been built. Here, the CarSim and TruckSim simulation and visualization environments play central roles. The vehicle simulator has several purposes. It demonstrates the operation of the predictive cruise control and provides the possibility to select the different design and operation parameters. The predictive speed control can be compared to conventional cruise control solutions in the online environment. In the second part of the chapter, some results from the real validation are also presented. The chapter also includes the architecture of realized control and the test results.

In the Appendix, further components of the traffic control are included, see Part IV. Chapter "Model-based robust control design" briefly summarizes the main steps of the robust control design from the modeling to the synthesis. Chapter "Maximum controlled invariants sets" presents both the theoretical background and the practical computation method of the control invariant sets.

References

Gáspár P, Szabó Z, Bokor J, Németh B (2017) Robust control design for active driver assistance systems: a linear-parameter-varying approach. Springer International Publishing, Heidelberg

Sename O, Gáspár P, Bokor J (2013) Robust control and linear parameter varying approaches. Springer, Heidelberg

Waymo (2017) Waymo safety report: on the road to fully self-driving. Technical report, Waymo. https://waymo.com/safetyreport/

Part I
Predictive Cruise Control

Chapter 2
Design of Predictive Cruise Control Using Road Information

Introduction and Motivation

As a result of growing global requirements, the automotive researchers are forced to develop flexible, reliable, and economical automotive systems which require less energy during the operation. Reducing fuel consumption is an important environmental and economic requirement for vehicle systems. Since the driveline system has an important role in the emission of the vehicle, the development of the longitudinal control systems is in the focus of the research and development of the vehicle industry. This chapter presents a method of how the required force and energy, and thus fuel consumption can be reduced when the external road information is taken into consideration during the journey. Moreover, it proposes the design of a new adaptive cruise control system, in which the longitudinal control incorporates the brake and traction forces in order to achieve the designed velocity profile.

The controllers applied in current adaptive cruise control systems are able to take into consideration only instantaneous effects of road conditions since they do not have information about the oncoming road sections. The cruise control systems automatically maintain a steady speed of a vehicle as set by the driver by setting the longitudinal control forces. In the following, road inclinations are taken into consideration in the design of the longitudinal control force. The aim in this calculation is to achieve a control force which is similar to the driver's requirement. For example, in front of the downhill slope, the driver can see the change in the curve of the road. Here the velocity of the vehicle increases, thus the control force of the vehicle before the slope can be reduced. As a result, at the beginning of the slope, the velocity of the vehicle decreases, thus it will increase from a lower value. Consequently, the brake system can be activated later or it may not be necessary to activate it at all. If the velocity in the next road section changes, it is possible to set the adequate control force. In the knowledge of the speed limits, it is also possible to save energy. Moreover, in the section of the road where a speed limit is imposed different strategies can be considered. Before the regulated section, the velocity can be reduced, therefore

P. Gáspár and B. Németh, *Predictive Cruise Control for Road Vehicles Using Road and Traffic Information*, Advances in Industrial Control,
https://doi.org/10.1007/978-3-030-04116-8_2

less energy is necessary for the vehicle. Using the idea of road slope and speed limit, fuel consumption and the energy required by the actuators can be reduced. By choosing the appropriate velocity according to the road and traffic information, the number of unnecessary accelerations and brakings and their durations can be significantly reduced.

In the vehicle, the most important longitudinal actuators are the engine, the transmission and the brake system. The engine is set at a particular revolution with corresponding consumption, torques, etc. If road conditions are known, the engine can be operated more efficiently throughout the entire journey. The transmission system has effects on the engine since it creates a connection between the engine and the wheels. The selected gear affects the operation of the engine. Hence, the engine and the transmission system must be handled together in a control system. Moreover, the unnecessarily frequent activation of the brake is undesirable because of the wear of the brake pad/disc and the loss in kinetic energy. The control of longitudinal dynamics requires the integration of these vehicle components, see e.g., Kiencke and Nielsen (2000), Trachtler (2004).

The method takes into consideration both the inclination of the road and the speed limits. Vehicles save energy at the change of road inclinations and at the same time keep compulsory speed limits. In addition, the tracking of the preceding vehicle is necessary to avoid a collision. If the preceding vehicle accelerates or decelerates, the tracking vehicle must strictly track the velocity within the speed limit. Thus, this method changes the speed according to the road and traffic conditions. At the same time, the efficiency of the transportation system as an important cost factor requires relatively steady speed. These requirements are in conflict and the trade-off among them can be achieved using different weights.

Several methods in which the road conditions are taken into consideration have already been proposed, see Ivarsson et al. (2009), Nouveliere et al. (2008), Németh and Gáspár (2010). The look-ahead control methods assume that information about the future disturbances to the controlled system is available. To find a compromise solution between fuel consumption and trip time leads to an optimization problem. The optimization was handled using a receding horizon control approach in Hellström et al. (2010), Passenberg et al. (2009). In another approach, the terrain and traffic flow were modeled stochastically using a Markov chain model in Kolmanovsky and Filev (2009, 2010). In Hellström et al. (2009), the approach was evaluated in real experiments where the road slope was estimated by the method in Sahlholm and Johansson (2009). The work Faris et al. (2011) classifies several modeling approaches for vehicle fuel consumption and emission, such as microscopic, mesoscopic, and macroscopic modeling methods. From the aspect of microscopic approach, models of vehicle dynamics are preferred in the paper. Alternative truck lane management strategies are evaluated in Rakha et al. (2006). The efficiency of this method is presented by different scenarios, which show that using these methods travel time, energy, and the emission of the vehicle can be reduced. Rakha et al. (2006, 1989) present modeling methods for the design of route guidance strategies and the reliable estimation of travel time. The preliminary results of the research are also published in Németh and Gáspár (2010).

The aim of the design method is to calculate the longitudinal forces by using an optimization method. The optimal solution is built into a closed-loop interconnection structure in which a robust controller is designed using a Linear Parameter Varying (LPV) method. In the LPV method uncertainties, disturbances and nonlinear properties of the system are also handled. The real physical inputs of the system (throttle, gear position, and brake pressure) are calculated using the longitudinal force required by velocity tracking. By choosing the appropriate velocity according to the road and traffic information, the number of unnecessary accelerations and brakings and their durations can be significantly reduced. The specific components such as actuators occur in the implementation task. An important feature of the method is that the optimization task and the implementation task are handled separately. Consequently, the method can be implemented in an ECU (electronic control unit) in practice.

2.1 Speed Design Based on Road Slopes and Weighting Factors

In this section, the road inclinations and speed limits are formalized in a control-oriented model. First, the road ahead of the vehicle is divided into several sections and reference velocities are selected for them. The rates of the inclinations of the road and those of the speed limits are assumed to be known at the endpoints of each section. Second, the road sections are qualified by different weights, which have an important role in control design. The appropriate selection of the weights creates a balance between the velocity of the vehicle and the effects of road conditions. The knowledge of the road inclinations is a necessary assumption for the calculation of the velocity signal. In practice, the slope of the road can be obtained in two ways: either a contour map which contains the level lines is used, or an estimation method is applied. In the former case, a map used in other navigation tasks can be extended with slope information. Several methods have been proposed for slope estimation. They use cameras, laser/inertial profilometers, differential GPS or a GPS/INS systems, see Bae et al. (2001), Labayrade et al. (2002), Hahn et al. (2004). An estimation method based on a vehicle model and Kalman filters was proposed by Lingman and Schmidtbauer (2002). The detection of a speed limit sign is usually based on a video camera.

The principle of the consideration of road conditions is the following. It is assumed that the vehicle travels in a segment from the initial point (beginning of the road section) to the first division point. The velocity at the initial point is predefined and it is called original velocity. The journey is carried out with constant longitudinal force. The dynamics of the vehicle is described between the initial and the first division points. An important question is how velocity should be selected at the initial point (called modified velocity) at which the reference velocity of the first point can be reached using a constant longitudinal force. The thought can be extended to the next segments and division points. In case of n number of segments, n equations are formalized between the first and the endpoints.

The number of segments is important. For example, in the case of flat roads, it is enough to use relatively few section points because the slopes of the sections do not change abruptly. In the case of undulating roads, it is necessary to use relatively large number of section points and shorter sections because it is assumed in the algorithm that the acceleration of the vehicle is constant between the section points. Thus, the road ahead of the vehicle is divided unevenly, which is consistent with the topography of the road.

2.1.1 Speeds at the Section Points Ahead of the Vehicle

The simplified model of the vehicle is shown in Fig. 2.1. The longitudinal movement of the vehicle is influenced by the traction force F_l as the control signal and the disturbance force F_d. The longitudinal force guarantees the acceleration of the vehicle

$$F_l = m\ddot{\xi} + F_d, \tag{2.1}$$

where m is the mass of the vehicle, $\ddot{\xi}$ is the acceleration, and the F_d disturbance is taken into consideration. The acceleration of the vehicle is the following:

$$\ddot{\xi} = (F_l - F_d)/m, \tag{2.2}$$

Several longitudinal disturbances influence the movement of the vehicle.

$$F_d = F_r + F_{aer} + G_x, \tag{2.3}$$

where F_r, F_{aer} and G_x are the rolling resistance, the aerodynamic force, and the weighting force, respectively. The rolling resistance is modeled by an empiric form

$$F_r = F_z f_0(1 + \kappa\dot{\xi}^2), \tag{2.4}$$

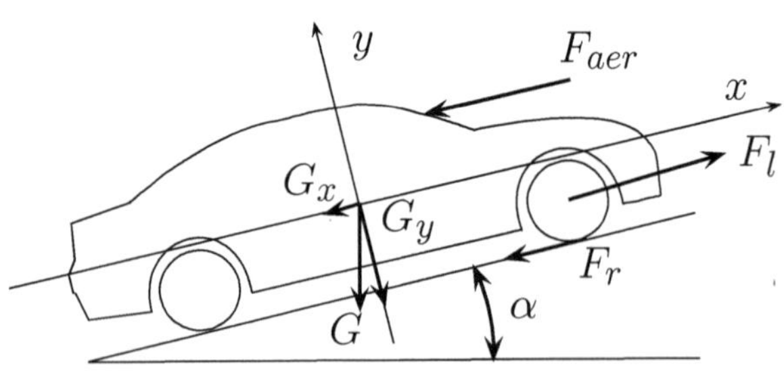

Fig. 2.1 Simplified vehicle model

where F_z is the vertical load of the wheel, f_0 and κ are empirical parameters depending on tyre and road conditions and $\dot{\xi}$ is the velocity of the vehicle, see Pacejka (2004). The aerodynamic force is formulated as

$$F_{aer} = 0.5 C_w \rho A_0 \dot{\xi}_{rel}^2, \tag{2.5}$$

where C_w is the drag coefficient, ρ is the density of air, A_0 is the reference area, $\dot{\xi}_{rel}$ is the velocity of vehicle relative to the air. In the following, a lull is assumed, i.e., $\dot{\xi}_{rel} = \dot{\xi}$. The longitudinal component of the weighting force is

$$G_x = mgsin\alpha, \tag{2.6}$$

where m is the mass of the vehicle and α is the angle of the slope.

The predicted course of the vehicle can be divided into sections using $n + 1$ number of points as Fig. 2.2 shown. Although between the points may be acceleration and declaration an average speed is used. Thus, the rate of accelerations of the vehicle is considered to be constant between these points.

It is assumed that the velocity in the starting point is the first reference velocity:

$$\dot{\xi}_0^2 = v_{ref,0}^2. \tag{2.7}$$

The displacement of the vehicle in the first section can be expressed by the velocity differences using simple kinematic equations is

$$s_1 = \frac{1}{2}(\dot{\xi}_1 + \dot{\xi}_0)t, \tag{2.8}$$

where $\dot{\xi}_0$ is the velocity of vehicle at the initial point, $\dot{\xi}_1$ is the velocity of vehicle at the first point, and s_1 is the distance between these points. Time t is expressed by the relationship between the acceleration and the relative velocity as follows:

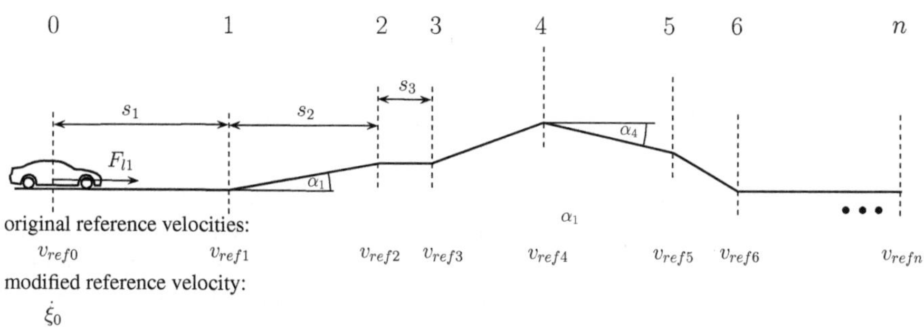

Fig. 2.2 Section points of the road ahead the vehicle

$$\ddot{\xi} = \frac{\dot{\xi}_1 - \dot{\xi}_0}{t}. \tag{2.9}$$

Consequently, the velocity of the first section point can be expressed by the initial velocity and the acceleration is as follows:

$$\dot{\xi}_1^2 = \dot{\xi}_0^2 + 2\ddot{\xi}s_1 = \dot{\xi}_0^2 + \frac{2}{m}s_1(F_{l1} - F_{d1}). \tag{2.10}$$

The velocity of the first section point $\dot{\xi}_1^2$ is defined as the reference velocity $v_{ref,1}^2 = \dot{\xi}_1^2$. This relationship also applies to the next road section:

$$\dot{\xi}_2^2 = \dot{\xi}_1^2 + 2\ddot{\xi}s_2. \tag{2.11}$$

The purpose of considering road conditions is to determine a control force by which the vehicle can travel along its way. It is important to emphasize that the longitudinal force F_{l1} is known only the first section $i = 1$ and not known further $i \in \{2, n\}$. At the calculation of the control force, it is assumed that additional longitudinal forces will not act on the vehicle, i.e., the longitudinal forces $F_{li},\ i \in \{2, n\}$ will not affect the next sections. At the same time, the disturbances from road slope are known ahead. Consequently, the velocity of the second section point $\dot{\xi}_2^2$, which is defined as the reference velocity $v_{ref,2}^2 = \dot{\xi}_2^2$ is as follows:

$$\dot{\xi}_2^2 = \dot{\xi}_0^2 + \frac{2}{m}(s_1 F_{l1} - s_1 F_{d1} - s_2 F_{d2}). \tag{2.12}$$

Similarly, the velocity of the vehicle can be formalized in the next n section points. Using this principle, a velocity-chain, which contains the required velocities along the way of the vehicle is constructed. At the calculation of the control force, it is assumed that additional longitudinal forces $F_{li}, i \in [2, n]$ will not affect the next sections. The velocities of vehicle are described at each section point of the road using similar expressions to (2.10). The velocity of the nth section point is the following:

$$\dot{\xi}_n^2 = \dot{\xi}_0^2 + \frac{2}{m}(s_1 F_{l1} - \sum_{i=1}^{n} s_i F_{di}) = v_{ref,n}^2. \tag{2.13}$$

It is also an important goal to track the momentary value of the velocity.

The F_{di} disturbance force can be divided into two parts: the first part is the force resistance from road slope $F_{di,r}$, while the second part $F_{di,o}$ contains all of the other resistances such as rolling resistance, aerodynamic forces, etc. The disturbance force is as follows:

$$F_{di} = F_{di,r} + F_{di,o}. \tag{2.14}$$

We assume that $F_{di,r}$ is known while $F_{di,o}$ is unknown. Using (2.6), $F_{di,r} = G_x$ depends on the mass of the vehicle and the angle of the slope α_i. When the control force F_{l1} is calculated, only $F_{d1,o}$ influences the vehicle of all of the unmeasured disturbances. In the control design, the effects of the unmeasured disturbances $F_{di,o}$, $i \in \{2, n\}$ are ignored. The consequence of this assumption is that the model does not contain all the information about the road disturbances, therefore it is necessary to design a robust speed controller. This controller can ignore the undesirable effects.

Consequently, as an example, the velocity of the nth section point is the following:

$$\dot{\xi}_n^2 = \dot{\xi}_0^2 + \frac{2}{m}(s_1 F_{l1} - \frac{2}{m} s_1 F_{d1,o} - \frac{2}{m} \sum_{i=1}^{n} s_i F_{di,r}. \tag{2.15}$$

Moreover, the equations of the vehicle at the section points are calculated in the following way:

$$\dot{\xi}_0^2 + \frac{2}{m} s_1 F_{l1} - \frac{2}{m} s_1 F_{d1,o} = v_{ref,1}^2 + \frac{2}{m} s_1 F_{d1,r} \tag{2.16}$$

$$\dot{\xi}_0^2 + \frac{2}{m} s_1 F_{l1} - \frac{2}{m} s_1 F_{d1,o} = v_{ref,2}^2 + \frac{2}{m} (s_1 F_{d1,r} + s_2 F_{d2,r}) \tag{2.17}$$

$$\vdots$$

$$\dot{\xi}_0^2 + \frac{2}{m} s_1 F_{l1} - \frac{2}{m} s_1 F_{d1,o} = v_{ref,n}^2 + \frac{2}{m} \sum_{i=1}^{n} s_i F_{di,r} \tag{2.18}$$

2.1.2 *Weighting Strategy*

In the next step, weights are introduced. Weight Q is applied to the current reference velocity, weights $\gamma_1, \gamma_2, ..., \gamma_n$ are applied to the reference velocities at the section points. Weight Q has an essential role: it determines the tracking requirement of the current reference velocity $v_{ref,0}$. Weights γ_i represent the rate of the road conditions ahead of the vehicle. The weights should sum up to one, i.e.,

$$\gamma_1 + \gamma_2 + ... + \gamma_n + Q = 1. \tag{2.19}$$

By increasing Q, the momentary velocity becomes more important while road conditions become less important.

Weight Q is applied in Eq. (2.7) and weights $\gamma_1, \gamma_2, ..., \gamma_n$ are applied in Eqs. (2.16)–(2.18), in order to achieve the relationship between the vehicle parameters and the weights

$$Q\dot{\xi}_0^2 = Qv_{ref,0}^2 \tag{2.20}$$

$$\gamma_1 \dot{\xi}_0^2 + \gamma_1 \frac{2}{m} s_1 F_{l1} - \gamma_1 \frac{2}{m} s_1 F_{d1,o} = \gamma_1 v_{ref,1}^2 + \gamma_1 \frac{2}{m} s_1 F_{d1,r} \tag{2.21}$$

$$\gamma_2 \dot{\xi}_0^2 + \gamma_2 \frac{2}{m} s_1 F_{l1} - \gamma_2 \frac{2}{m} s_1 F_{d1,o} = \gamma_2 v_{ref,2}^2 + \gamma_2 \frac{2}{m} (s_1 F_{d1,r} + \gamma_2 s_2 F_{d2,r}) \tag{2.22}$$

$$\vdots$$

$$\gamma_n \dot{\xi}_0^2 + \gamma_n \frac{2}{m} s_1 F_{l1} - \gamma_n \frac{2}{m} s_1 F_{d1,o} = \gamma_n v_{ref,n}^2 + \gamma_n \frac{2}{m} \sum_{i=1}^{n} s_i F_{di,r} \tag{2.23}$$

By taking the weights into consideration when Eqs. (2.20)–(2.23) are summarized, the following formula is yielded:

$$\begin{aligned}
&(Q + \gamma_1 + \gamma_2 + \ldots + \gamma_n)\dot{\xi}_0^2 + \frac{2}{m} s_1 (\gamma_1 + \gamma_2 + \ldots + \gamma_n) F_{l1} - \frac{2}{m} s_1 (\gamma_1 + \gamma_2 + \ldots + \gamma_n) F_{d1,o} \\
&\quad = Q v_{ref,0}^2 + (\gamma_1 v_{ref,1}^2 + \gamma_2 v_{ref,2}^2 + \ldots + \gamma_n v_{ref,n}^2) \\
&\quad + \frac{2}{m} s_1 F_{d1,r} (\gamma_1 + \gamma_2 + \ldots + \gamma_n) + \frac{2}{m} s_2 F_{d2,r} (\gamma_2 + \ldots + \gamma_n) + \ldots + \frac{2}{m} s_n F_{dn,r} \gamma_n
\end{aligned} \tag{2.24}$$

Applying (2.19) to the Eq. (2.24), the following equation is achieved:

$$\begin{aligned}
&\dot{\xi}_0^2 + \frac{2}{m} s_1 (1 - Q) F_{l1} - \frac{2}{m} s_1 (1 - Q) F_{d1,o} \\
&\quad = Q v_{ref,0}^2 + \sum_{i=1}^{n} \gamma_i v_{ref,i}^2 \\
&\quad + \frac{2}{m} s_1 F_{d1,r} \sum_{i=1}^{n} \gamma_i + \frac{2}{m} s_2 F_{d2,r} \sum_{i=2}^{n} \gamma_i + \ldots + \frac{2}{m} s_n F_{dn,r} \sum_{i=n}^{n} \gamma_i
\end{aligned} \tag{2.25}$$

The final equation is the following:

$$\dot{\xi}_0^2 + \frac{2}{m} s_1 (1 - Q) F_{l1} - \frac{2}{m} s_1 (1 - Q) F_{d1,o} = \vartheta, \tag{2.26}$$

where the right-hand side, i.e., the value ϑ depends on the road slopes, the reference velocities and the weights

$$\vartheta = Q v_{ref,0}^2 + \sum_{i=1}^{n} \gamma_i v_{ref,i}^2 + \frac{2}{m} \sum_{i=1}^{n} s_i F_{di,r} \sum_{j=i}^{n} \gamma_j. \tag{2.27}$$

Note that weights have an important role both in the further analysis and synthesis. By making an appropriate selection of the weights, the importance of the road condition is taken into consideration. For example, when

$$Q = 1, \quad \gamma_i = 0, \ i \in [1, n] \tag{2.28}$$

the control exercise is simplified to a cruise control problem without any road conditions. When equivalent weights are used the road conditions are considered with the same importance, i.e.,

$$Q = \gamma_1 = \gamma_2 = \ldots = \gamma_n. \tag{2.29}$$

The optimal determination of the weights has an important role, i.e., to achieve a balance between the current velocity and the effect of the road slope. Consequently, a balance between the velocity and the economy parameters of the vehicle is formalized.

In the final step, a control-oriented vehicle model, in which reference velocities and weights are taken into consideration, is constructed. The momentary acceleration of the vehicle is expressed in the following way:

$$\ddot{\xi}_0 = \frac{F_{l1} - F_{d1,o} - F_{d1,r}}{m}, \tag{2.30}$$

where $F_{d1,r} = mg \sin\alpha$. Thus,

$$\frac{F_{l1} - F_{d1,o}}{m} = \ddot{\xi}_0 + g \sin\alpha. \tag{2.31}$$

Equation (2.26) is rearranged:

$$\dot{\xi}_0 = \lambda, \tag{2.32}$$

where the parameter λ is calculated in the following way based on the Eq. (2.26) and the designed ϑ according to (2.27):

$$\lambda = \sqrt{\vartheta - 2s_1(1 - Q)(\ddot{\xi}_0 + gsin\alpha)}, \tag{2.33}$$

Consequently, the road conditions can be considered by velocity tracking. The momentary velocity of vehicle $\dot{\xi}$ should be equal to parameter λ, which contains the road information. The calculation of λ requires the measurement of the longitudinal acceleration $\ddot{\xi}_0$.

2.2 Optimization of the Vehicle Cruise Control

Equation (2.26) shows that the modified velocity $\dot{\xi}_0$ depends on the weights (Q and γ_i). In the following, the longitudinal force is expressed by the weights. From (2.26), the following expression can be created:

$$
\begin{aligned}
F_{l1} = \frac{m}{2s_1(1-Q)}[& Qv_{ref,0}^2 + \gamma_1 v_{ref,1}^2 + \gamma_2 v_{ref,2}^2 + \ldots + \gamma_n v_{ref,n}^2 + \\
& + \frac{2}{m}s_1 F_{d1,r}\gamma_1 + \frac{2}{m}s_1 F_{d1,r}\gamma_2 \ldots \frac{2}{m}s_1 F_{d1,r}\gamma_n + \\
& + \frac{2}{m}s_2 F_{d2,r}\gamma_2 + \frac{2}{m}s_2 F_{d2,r}\gamma_3 \ldots \frac{2}{m}s_2 F_{d2,r}\gamma_n + \\
& + \ldots + \\
& + \frac{2}{m}s_n F_{dn,r}\gamma_n - \dot{\xi}_0^2 + \frac{2}{m}s_1(1-Q)F_{d1,o}]
\end{aligned} \tag{2.34}
$$

The longitudinal force can be expressed by the weights in the following way:

$$
F_{l1} = \beta_0(Q) + \beta_1(Q)\gamma_1 + \beta_2(Q)\gamma_2 + \ldots + \beta_n(Q)\gamma_n + K, \tag{2.35}
$$

where the members can be expressed by the weights Q and γ_i using the comparison (2.34) and (2.35):

$$
\beta_0(Q) = \frac{m}{2s_1(1-Q)}(Qv_{ref,0}^2 - \dot{\xi}_0^2 \tag{2.36}
$$

$$
\beta_i(Q)\gamma_i = \frac{m}{2s_1(1-Q)}\left(v_{ref,i}^2 + \frac{2}{m}\sum_{j=1}^{i} s_j F_{dj,r}\right)\gamma_i \tag{2.37}
$$

$$
K = F_{d1,o} \tag{2.38}
$$

In this section, the task is to find an optimal selection of the weights in such a way that both the minimization of control force and the traveling time are taken into consideration.

The vehicle cruise control problem can be divided into two optimization problems in the following forms:

Optimization 1: The longitudinal control force must be minimized, i.e., $|F_{l1}| \rightarrow Min!$. Instead, in practice, the

$$
F_{l1}^2 \rightarrow Min! \tag{2.39}
$$

optimization is used because of the simpler numerical computation.

Optimization 2: The difference between reference velocity and modified velocity must be minimized, i.e.,

$$
|v_{ref,0} - \dot{\xi}_0| \rightarrow Min! \tag{2.40}
$$

2.2.1 Handling the Optimization Criteria

The two optimization criteria lead to different optimal solutions. In the first criterion, the predicted road inclinations and speed limits are taken into consideration by using appropriately chosen weights $\hat{Q}$, $\hat{\gamma}_i$. At the same time, the second criterion is optimal if the predicted information is neglected. In the latter case, the prediction weights are noted by $\breve{Q}$, $\breve{\gamma}_i$.

The first criterion (Optimization 1) is met by the formalization of a quadratic optimization problem. It leads to the following form:

$$\hat{F}_{l1}^2(\hat{Q}, \hat{\gamma}_i) = (\beta_0(\hat{Q}) + \beta_1(\hat{Q})\hat{\gamma}_1 + \beta_2(\hat{Q})\hat{\gamma}_2 + \ldots + \beta_n(\hat{Q})\hat{\gamma}_n)^2 \tag{2.41}$$

with the following constrains:

$$\bar{\gamma}_1 + \bar{\gamma}_2 + ... + \bar{\gamma}_n + \bar{Q} = 1 \tag{2.42}$$

$$0 \leq \bar{Q}, \bar{\gamma}_i \leq 1 \tag{2.43}$$

This task is nonlinear because of the prediction weights. In the first optimization, the prediction weight $\hat{Q}$ is fixed, and this fixed $\hat{Q}$ is modified from 0 to 1. The optimization task is formulated in the following matrix form. Since the optimization task is linear in the sense of prediction weights $\hat{\gamma}_i$, (2.41) is formulated in the following matrix form:

$$\hat{F}_{l1}^2(\Gamma) = \frac{1}{2}\Gamma^T \Phi \Gamma - \kappa \Gamma, \tag{2.44}$$

where the matrix Γ is $\Gamma = \begin{bmatrix} \hat{\gamma}_1 & \cdots & \hat{\gamma}_i & \cdots & \hat{\gamma}_n \end{bmatrix}^T$ and the matrix Φ comes from the rearrangement of (2.41). Thus, the problem leads to a quadratic programming task.

The condition analysis is crucial, since it is related to the appropriation of the numerical solution. For example, with a flat road and constant velocity regulations, the values of $\beta_i(\hat{Q})$ $(1 \leq i \leq n)$ are relatively the same. Since, in this case, the elements of matrix Φ are equal to each other, the matrix Φ is singular. Consequently, the computation of Φ^{-1} is difficult or impossible, and the condition number of Φ is very high. Since in practice several similar situations can be obtained, a numerical algorithm should be applied which is able to handle the poor conditioning system. The Levenberg–Marquardt algorithm is able to handle the deficiency of the conditioning system, see Marquardt (1963). In this method, the original matrix Φ is increased by an identity matrix I multiplied by a small number $(\delta > 0)$: $\hat{\Phi} = \Phi + \delta I$. By the Levenberg–Marquardt algorithm, the condition number of Φ can be reduced significantly, which helps solve the optimization task. In the next step, the quadratic optimization task is derived: $\Gamma = -\hat{\Phi}^{-1}\kappa$.

The optimization task is solved only on a bounded range of the weights: $0 \leq Q, \gamma_i \leq 1$ and $Q + \sum \gamma_i = 1$. The solution of this task is difficult and it requires a great deal of computation besides decreasing the condition number of Φ. In practice, the numerical computations result in optimal weights, which change very sharply as a jump signal. In order to avoid this phenomenon, the weights are filtered by low-pass filters to obtain smooth signals.

The second criterion (Optimization 2) is also taken into consideration. The optimal solution can be determined in a relatively easy way since the vehicle tracks the predefined velocity if the predicted road conditions are not considered. Consequently, the optimal solution is achieved by selecting the prediction weights in the following way: $\breve{Q} = 1$ and $\breve{\gamma}_i = 0, i \in [1, n]$.

Finally, a balance between the two performances must be achieved, which is based on a tuning of the designed prediction weights. The first criterion is met by selecting prediction weights $\bar{Q}$, $\bar{\gamma}_i$. The second performance is met by selecting constant prediction weights

$$\breve{Q} = 1, \breve{\gamma}_i = 0, \; i \in [1, n]. \tag{2.45}$$

2.2.2 *Trade-Off Between the Optimization Criteria*

Several methods can be applied in this task. In the proposed method, two further performance weights, i.e., R_1 and R_2 are introduced. The performance weight R_1 $(0 \leq R_1 \leq 1)$ is related to the importance of the minimization of the longitudinal control force F_{l1} (Optimization 1) while the performance weight R_2 $(0 \leq R_2 \leq 1)$ is related to the minimization of $|v_{ref,0} - \dot{\xi}_0|$ (Optimization 2). There is a constraint according to the performance weights

$$R_1 + R_2 = 1. \tag{2.46}$$

Thus, the performance weights, which guarantee balance between optimization tasks, are calculated in the following expressions:

$$Q = R_1\bar{Q} + R_2\breve{Q} = R_1\bar{Q} + R_2 = 1 - R_1(1 - \bar{Q}) \tag{2.47}$$

$$\gamma_1 = R_1\bar{\gamma}_1 + R_2\breve{\gamma}_1 = R_1\bar{\gamma}_1 \tag{2.48}$$

$$\vdots$$

$$\gamma_n = R_1\bar{\gamma}_n + R_2\breve{\gamma}_n = R_1\bar{\gamma}_n \tag{2.49}$$

The equations show that the prediction weights depend on R_1. Using the (2.36) and (2.37) $\beta_i(Q)$ can be expressed:

$$
\begin{aligned}
\beta_0(Q) &= \frac{m}{2s_1 R_1(1-\bar{Q})}\left((1-R_1+R_1\bar{Q})v_{ref,0}^2 - \dot{\xi}_0^2\right) \\
&= \frac{m}{2s_1 R_1(1-\bar{Q})}(v_{ref,0}^2 - \dot{\xi}_0^2) - \frac{m}{2s_1}v_{ref,0}^2 \\
&= \beta_0(\bar{Q}) - \frac{m}{2s_1}v_{ref,0}^2
\end{aligned} \tag{2.50}
$$

$$
\begin{aligned}
\beta_i(Q)\gamma_i &= \frac{m}{2s_1 R_1(1-\bar{Q})}\left(v_{ref,i}^2 + \frac{2}{m}\sum_{j=1}^{i} s_j F_{dj,r}\right) R_1\bar{\gamma}_i \\
&= \beta_i(\bar{Q})\bar{\gamma}_i
\end{aligned} \tag{2.51}
$$

Normally, drivers set weight R_1 based on their goals and requirements, thus they create a balance between energy saving and traveling time. Based on the calculated performance weights the modified velocity can be determined using (2.26). In practice, the numerical computations result in optimal prediction weights, which may change very sharply as a jump signal. In order to avoid this phenomenon, the prediction weights are filtered by low-pass filters to get smooth signals.

The optimal momentary speed of the vehicle is approximated by

$$
\dot{\xi}_0 = \lambda_{opt}, \tag{2.52}
$$

where the parameter λ_{opt} is calculated in the following way based on the designed ϑ:

$$
\lambda_{opt} = \sqrt{\vartheta - 2s_1(R_1(1-\bar{Q}))(\ddot{\xi}_0 + g sin\alpha)} \tag{2.53}
$$

and

$$
\vartheta = v_{ref,0}^2 - R_1(1-\bar{Q})v_{ref,0}^2 + R_1\sum_{i=1}^{n}\bar{\gamma}_i v_{ref,i}^2 + R_1\frac{2}{m}\sum_{i=1}^{n} s_i F_{di,r}\sum_{j=i}^{n}\bar{\gamma}_j. \tag{2.54}
$$

2.2.3 *Handling Traveling Time*

Since traveling time has great importance for both individual drivers and transportation companies, keeping a scheduled traveling time has a big effect on the acceptance of such autonomous systems. Hence, managing traveling time is a key element of the proposed method, which is guaranteed by the proper selection of the tuning parameter R_1. Hence, in order to keep the desired traveling time constraints in the optimization of R_1 is needed. For this purpose, the actual and predicted future motion of the controlled vehicle is necessary. Therefore, the vehicle speed on the look-ahead road sections must be estimated. The velocities of the ith section points is the following:

$$\dot{\xi}_i^2 = \dot{\xi}_0^2 + \frac{2}{m} s_1 F_{l1} - \frac{2}{m} F_{d1,o} - \frac{2}{m} \sum_{j=1}^{n} s_j F_{dj}, \quad i = \{1, ..., n\} \tag{2.55}$$

where $\dot{\xi}_0$ depends on the selected weighting gain R_1 according to (2.33) and (2.54). In the calculation, the weights $\bar{Q}$ and $\bar{\gamma}$ received from the optimization procedures are also included.

Consideration of traveling time in cruise control is a crucial task, especially for commercial vehicles. The design and implement of optimal speed is performed if the desired traveling time does not exceed the required traveling time. The required traveling time is denoted by Δt_{max}. However, if the time requirement cannot be accomplished, the speed strategy must be overwritten. In this case, the maximum speed should be applied.

In the following, a prediction for the desired traveling time is presented. Assuming constant accelerations on the forward road sections, it is possible to predict the traveling time between section points as follows:

$$\Delta t_1 = \frac{2s_1}{\dot{\xi}_1 + \lambda}, \quad i = 1 \tag{2.56}$$

$$\Delta t_i = \frac{2s_i}{\dot{\xi}_i + \dot{\xi}_{i-1}}, \quad i = 2 \ldots n \tag{2.57}$$

The overall travel time on the look-ahead horizon can be calculated by summing up travel times between section points:

$$\Delta t = \sum_{i=1}^{n} \Delta t_i \tag{2.58}$$

According to (2.55), the predicted traveling time is a function of R_1.

Then, the desired traveling time Δt_{min} is calculated by taking speed limits into consideration. It must be compared to both the predicted traveling time Δt and the required traveling time Δt_{max}. For this purpose, based on the speed limits $v_{ref,i}$ the traveling time is computed for each road section as follows:

$$\Delta t_{i,min} = \frac{2s_i}{v_{ref,i} + v_{ref,i-1}}, \quad i = 1 \ldots n. \tag{2.59}$$

Hence, the overall minimum traveling time can be given as sum of the minimum time values:

$$\Delta t_{min} = \sum_{i=1}^{n} \Delta t_{i,min}. \tag{2.60}$$

The computed Δt_{min} results in the minimum traveling time of the vehicle on the horizon.

Next, the differences among the predicted traveling time Δt, the minimum traveling time Δt_{min}, and the required traveling time Δt_{max} are analyzed. From these, Δt depends on the weighting parameter R_1. If $R_1 = 0$, i.e., the optimization focuses only on the traveling time and theoretically the difference $\Delta t - \Delta t_{min} = 0$. If $R_1 > 0$ then the difference will be positive since the predicted traveling time increases: $\Delta t - \Delta t_{min} > 0$. It results in a time delay on the traveling horizon.

Moreover, it is possible to limit the time delay on the entire route by selecting an appropriate weight R_1. This design task is important if the predicted traveling time exceeds the required traveling time: $\Delta t - \Delta t_{max} > 0$. A maximum value of the acceptable time delay t_{max} is introduced. It is defined by the driver or a fleet management system based on the transportation requirement. In the selection t_{max}, both the minimum traveling time Δt_{min} and the required traveling time Δt_{max} are taken into consideration.

The purpose of the speed design is to achieve a balance between longitudinal energy and traveling time. Thus, the design usually leads to the positive value of $R_1 > 0$. Besides the overall delay of the vehicle must be checked, i.e., it must be smaller or equal to t_{max}.

$$\Delta t - \Delta t_{min} \leq t_{max}. \tag{2.61}$$

Note that the constraint depends on R_1 according to (2.33) and (2.54). The length of the acceptable time delay in (2.61) has a high impact because it influences Δt and Δt_{min}. The longer time horizon is set, the longer road section can be considered. Certainly, if (2.61) does not fulfilled, the parameter $R_1 = 0$ for the remaining road sections must be selected.

2.3 LPV Control Design Method

2.3.1 Control-Oriented LPV Modeling

The velocity tracking requires a controller which generates the longitudinal force. In this section, a high-level controller which calculates the longitudinal force is required. Note that the realization of the longitudinal force requires another low-level controller which sets the throttle angle and the gear position in the case of driving, or brake pressure in the case of braking. The longitudinal dynamics of the vehicle is formalized in the following form: $m\ddot{\xi}_0 = F_{l1} - F_{d1}$.. Both the driveline and braking systems have delays in their operations. The delay is caused by different factors such as the inertia of the driveline, the burning processes and injection, the turbo lag at driving, and the inertia of the hydraulic (or pneumatic) component in the braking system. The actuator dynamics is approximated by a first-order system Swaroop and Hedrick (1996):

$$\dot{F}_{l1} = \frac{1}{\tau}(\tilde{F}_{l1} - F_{l1}), \tag{2.62}$$

where F_{l1} is the realized force, $\tilde{F}_{l1}$ is the desired force of the vehicle, and τ is the delay of the system. Moreover, the delay parameter differs at driving (τ_d) and at braking (τ_b). More precisely, at braking, the delay is less than at driving, i.e., $\tau_d > \tau_b$. Therefore, the delay parameter is a varying component in the system.

The equations of the longitudinal dynamics and actuator dynamics are transformed into the following state-space representation form:

$$\dot{x} = Ax + B_1 F_{d1} + B_2 \tilde{F}_{l1}, \tag{2.63}$$

where the state vector is $x = \begin{bmatrix}\dot{\xi}_0 & F_{l1}\end{bmatrix}^T$ and the matrices in the state-space representation are

$$A = \begin{bmatrix} 0 & 1/m \\ 0 & -1/\tau \end{bmatrix}, \quad B_1 = \begin{bmatrix} -1/m \\ 0 \end{bmatrix}, \quad B_2 = \begin{bmatrix} 0 \\ 1/\tau \end{bmatrix}.$$

In the state-space description, the operations of the driving and the braking are handled simultaneously. Since $\tau_d > \tau_b$, the model is able to separate the driving and braking cases depending on τ.

The LPV model is based on the possibility of rewriting the plant in a form in which time-varying terms can be hidden with suitably defined scheduling variables. The LPV modeling approaches allow us to take the time-varying effects into consideration in the state-space description. Furthermore, this state space representation of the LPV model is valid in the entire operating region of interest. The advantage of LPV methods is that the controller meets robust stability and performance demands in the entire operational interval since the controller is able to adapt to the current operational conditions.

Selecting the scheduling variable τ, the model can be transformed into an LPV model:

$$\dot{x} = A(\rho)x + B_1 F_{d1} + B_2(\rho) \tilde{F}_{l1}, \tag{2.64}$$

where ρ is the scheduling variable:

$$\rho = \begin{cases} \tau_d & \text{in driving case} \\ \tau_b & \text{in braking case} \end{cases} \tag{2.65}$$

The aim of tracking is to ensure that the system output follows a reference value with an acceptable error, which is the performance of the system. The explicit mathematical description of the optimization problem is as follows:

$$(\dot{\xi}_0 - \lambda) \longrightarrow Min!, \tag{2.66}$$

where parameter λ is the reference value. In the velocity tracking problem, $z = \dot{\xi}_0 - \lambda$ is the performance output.

2.3.2 LPV-Based Control Design

The closed-loop interconnection structure, which includes the feedback structure of the model P and controller K is shown in Fig. 2.3.

The control design is based on a weighting strategy. The purpose of weighting function W_p is to define the performance specifications of the control system, i.e., the velocity of the vehicle must ensure the tracking of the reference signal with an acceptable error. They can be considered as penalty functions, i.e., the weights should be large where small signals are desired and small where large performance outputs can be tolerated. The formalized vehicle model approximates the driveline/braking system with a rigid body model. In case of real vehicles both driveline and braking systems have torsional or longitudinal vibrations. The natural frequencies of these

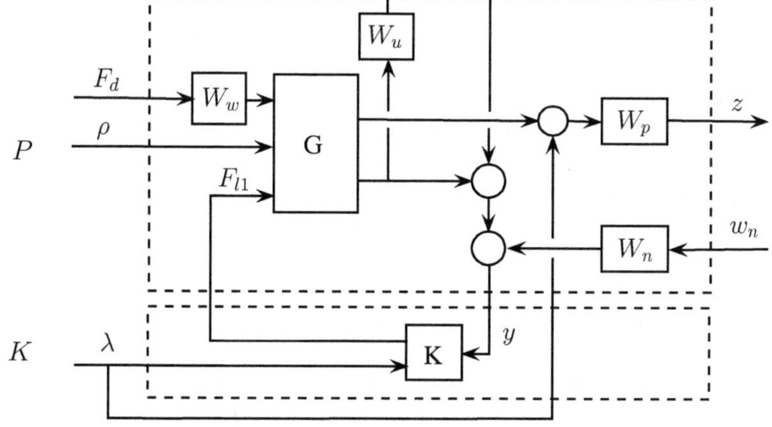

Fig. 2.3 Closed-loop interconnection structure

effects increase on higher frequencies. The weighting function W_p is selected as $W_p = \alpha/(Ts + 1)$, where α and T are constants.

The purpose of the weighting function W_n is to reflect the sensor noise, while W_w represents the effect of longitudinal disturbances. In the modeling, an unstructured uncertainty is modeled by connecting an unknown but bounded perturbation block ($\|\Delta\|_\infty < 1$) to the plant. The unstructured perturbation is connected to the plant in an output multiplicative structure. The magnitude of multiplicative uncertainty is handled by a weighting function W_u. The weighting functions W_u, W_w, and W_n are selected in linear and proportional forms. Note that although weighting functions are formalized in the frequency domain, their state-space representation forms are applied in the weighting strategy and in the control design.

In the design of the control system, the quadratic LPV performance problem is to choose the parameter-varying controller in such a way that the resulting closed-loop system is quadratically stable and, with zero initial conditions, the induced $\mathscr{L}_2$ norm from w to z is less than γ.

$$\|M(\rho)\|_\infty = \inf_K \sup_{\rho \in \mathscr{F}_\mathscr{P}} \sup_{\|w\|_2 \neq 0, w \in \mathscr{L}_2} \frac{\|z\|_2}{\|w\|_2} < \gamma \tag{2.67}$$

The existence of a controller that solves the quadratic LPV γ-performance problem can be expressed as the feasibility of a set of Linear Matrix Inequalities (LMIs), which can be solved numerically.

The LPV systems in early applications were based on a single Lyapunov function (SLF) approach, in which the variation of the scheduling variables can be arbitrarily fast. This work has been extended to analysis and synthesis by incorporating a parameter-dependent Lyapunov function, see Bokor and Balas (2005), Packard and Balas (1997), Wu et al. (1996). The incorporation of a parameter-dependent Lyapunov function implies a potentially less conservative approach by addressing limitations on the rate of change of the parameters ρ. However, the control design leads to infinite dimensional convex feasibility conditions. These conditions can, in general, only be obtained approximately, by selecting grid points from the whole set, thus it is converted into finite-dimensional LMIs.

Note that if parameter-dependent Lyapunov functions are used, the controller designed depends explicitly on $\dot{\rho}$. Thus, in order to construct a parameter-dependent controller, both ρ and $\dot{\rho}$ must be measured or available. When $\dot{\rho}$ is not measured in practice, a suitable extrapolation algorithm must be used to achieve an estimation of the parameter $\dot{\rho}$. The disadvantage of this approach is that the sources of the scheduling variables are not independent. Balas et al. proposed a possible method to perform a ρ-dependent change of variables to remove $\dot{\rho}$ dependence, see Packard and Balas (1997).

When the LPV controller has been synthesized, the relation between the state, or output, and the parameter $\rho = \sigma(x)$ is used in the LPV controller, such that a nonlinear controller is obtained. Note that it is assumed that $\sigma(x)$ is measured or depends only on measured signals. According to the properties of the LPV description, the LPV system with $\rho = \sigma(x)$ is equal to the nonlinear model. The realization of an LPV controller poses a problem, which must be handled. The control design is performed in continuous time, in which it is assumed that the scheduling variable is known in continuous time. However, ρ is measured only at sampling times. Instead of getting a fixed dependence of system matrices on ρ, the matrices are only known at a discrete ρ values. The suitable sampling time must be selected according to the physical system; however, the real sampling time is modified by the implementation possibilities. Thus, the determination of the parameters during the intervals between sampling times is a difficult theoretical problem. A simple procedure applied in practice uses a zero-order hold method between sampling times. A better solution of the approximation is based on polynomial or rational functions through curve fitting.

2.3.3 Stability Analysis of the Closed-Loop System

The computation of λ depends on the velocity $\dot{\xi}_0$ and the acceleration $\ddot{\xi}_0$ signals of the vehicles. These signals are used in the computation of the reference velocity signal, which can be considered as a feedback in the closed-loop system. The computed reference signal influences the stability of the controlled system. This stability problem is illustrated in Fig. 2.4. Figure 2.4a shows that in a conventional cruise control problem the reference speed can be considered as an exogenous signal. However, in the predictive look-ahead control problem the reference speed depends on several vehicle states, e.g., the acceleration and the current speed, see Fig. 2.4b. Therefore, the vehicle states influence the stability of the entire system.

Since the relationship between the vehicle parameters and λ is nonlinear, a simplification is used in the formulation, which requires the modeling of the reference signal generator. In the following a transfer function $G_{gen} = \left[G_{genA}, \; G_{genB}\right]$ is introduced. G_{genA} represents the relationship between the actual speed $\dot{\xi}_{0,j}$ and the reference signal λ, while G_{genB} is the transfer function between the acceleration and λ. G_{gen} is identified in an autoregressive with exogenous input structure (ARX), see, e.g., Ljung (1999):

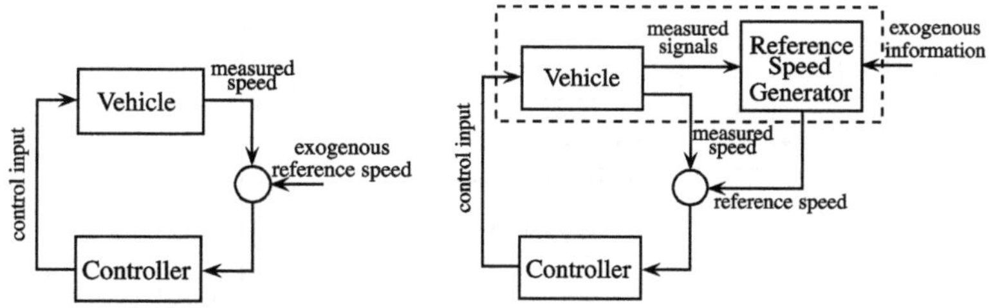

(a) Conventional problem (b) Look-ahead problem

Fig. 2.4 Problem setup

$$y_j(q) = \frac{B_{j,1}(q)}{A_j(q)} u_{j,1}(q) + \frac{B_{j,2}(q)}{A_j(q)} u_{j,2}(q) + e(q), \tag{2.68}$$

where $B_{j,i}(q)/A_j(q), i \in 1, 2$ functions are the discrete time models of the continuous systems G_{genA}, G_{genB}, moreover, $y(q)$ and $u_{j,i}(q)$ are discrete signals, which are sampled from $\lambda_j(t)$ and $\dot{\xi}_{0,j}(t)$, while $e(q)$ is noise.

Since the resulting systems vary in their parameters, it is required to consider their variations in a robust modeling structure. The structure of the reference speed generator is shown in Fig. 2.5. The variation of G_{gen} is considered in an additive uncertainty structure, where $G_{genA} = G_{gen1} + W_{u1}$ and $G_{genB} = G_{gen2} + W_{u1}$. G_{gen1}, G_{gen2} are the nominal systems, and W_{u1}, W_{u2} represent the additive uncertainty structure. Moreover, G_{filt} constant transfer function is a low-pass filter, which smoothes the computed λ reference signal. Note that this function must be used because the measured signals for the computation of λ are sampled.

Thus, the consideration of the reference signal generator means that it is necessary to incorporate it in the closed-loop interconnection structure of the cruise control

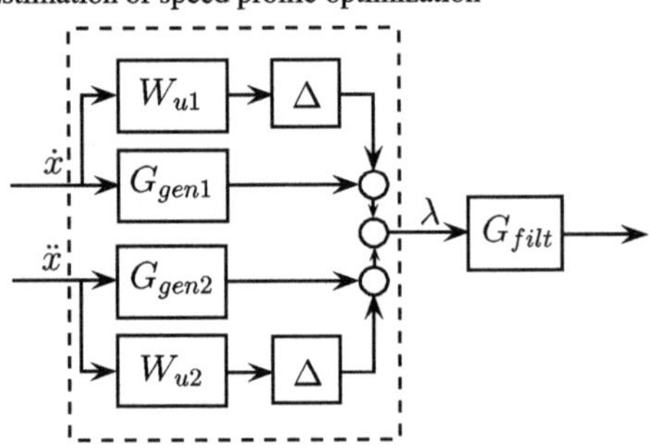

Fig. 2.5 Structure of the reference speed generator

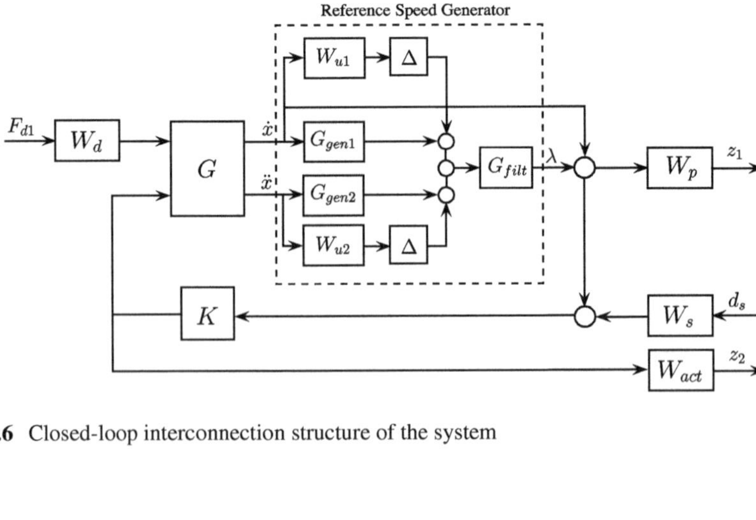

Fig. 2.6 Closed-loop interconnection structure of the system

design, as it is found in Fig. 2.6. In this way, the robust control is able to guarantee the stability of the vehicle together with the predictive speed selection strategy. It results in that the robustness criterion Zhou et al. (1996)

$$\left\| T_{z_\infty F_{d,fb}}(s) \right\|_\infty < \gamma, \tag{2.69}$$

for a given real positive value γ must be guaranteed, where $T_{z_\infty d}(s)$ is the transfer function from the disturbances to the performances.

2.3.4 Architecture of the Speed Profile Implementation

The architecture of the proposed computation algorithm is illustrated in Fig. 2.7. It shows that the vehicle receives information about the motion of preceding vehicle (distance and speed) and the follower vehicle (distance, speed, and acceleration). Moreover, the road inclination α_i and speed limit $v_{ref,i}$ data are also known. Since the route of the vehicle is considered as known, the loading of this information requires the measurement of the current position, e.g., using GPS.

A survey of the future communication possibilities in automotive and traffic control is provided by Ebnre and Hermann (2001). In order to detect potentially hazardous situations, a computer vision-based approach to tracking surrounding vehicles and estimating their trajectories were presented by Nuevo et al. (2010). An extension of adaptive cruise control with traffic information considering vehicle-to-roadside and vehicle-to-vehicle communication was proposed in Festag et al. (2008). To pre-

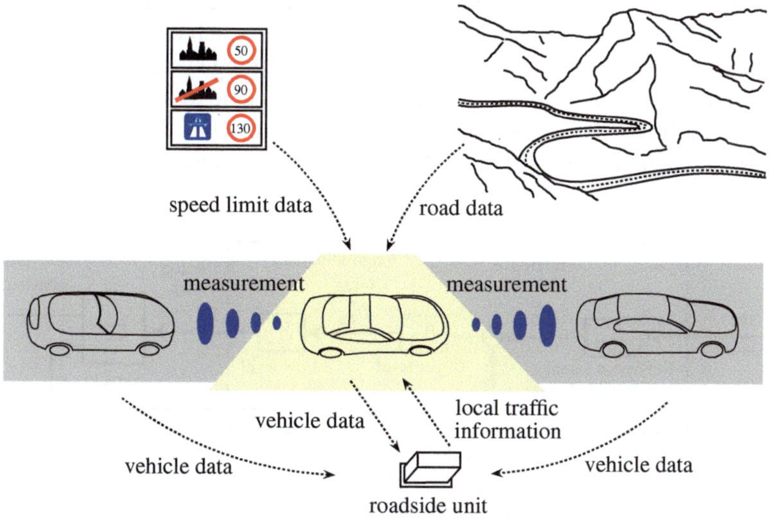

Fig. 2.7 Communication and information flow

vent emergencies and accidents, vehicle-to-vehicle communication and vehicle-to-roadside sensor communication were combined by Festag et al. (2008).

2.3.5 Architecture of the Control System

The control system can be realized in three steps as Fig. 2.8 shows.

Step 1: The aim of the first step is the computation of the reference velocity, see Eq. (2.32). The results of this computation are the weights and the modified velocity which must be tracked by the vehicle. The reference velocity is realized in two steps.

Step 2: In the second step, the longitudinal control force of the vehicle (F_{l1}) is designed. The role of the high-level controller K_1 is to calculate this required longitudinal force.

Step 3: In the third step, the real physical inputs of the system, such as the throttle, the gear position, and the brake pressure are generated by the low-level controller K_2.

In the proposed method, the steps are separated from each other. The reference velocity signal generator can be added to the upper-lower Adaptive Cruise Control (ACC) system. It is possible to design a reference signal generator unit almost individually and to attach it to the ACC system. Thus, the reference signal unit can be designed and produced independently from automobile suppliers, only a few vehicle data are needed. The independent implementation possibility is an important advantage in practice.

The purpose of the control design is to calculate the necessary longitudinal control force. It focuses on this high-level controller K, which calculates positive and negative forces as well, therefore the traction and braking system are also actuated. It is assumed that there is another controller which is able to track the control force as a required force. This low-level Electronic Control Unit (ECU) transforms the longitudinal force into a real physical parameter of the actuator. The design of this

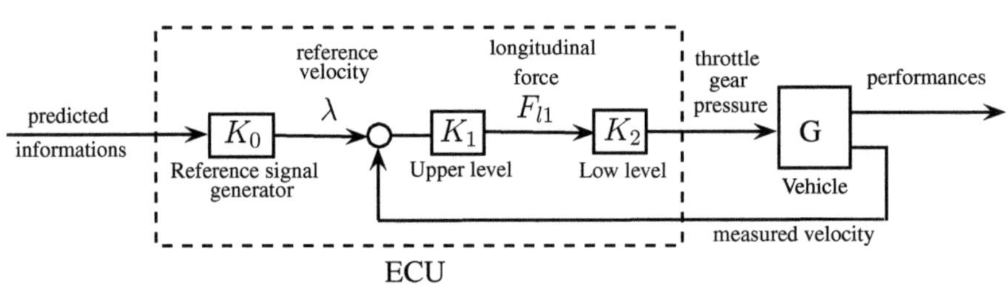

Fig. 2.8 Implementation of the controlled system

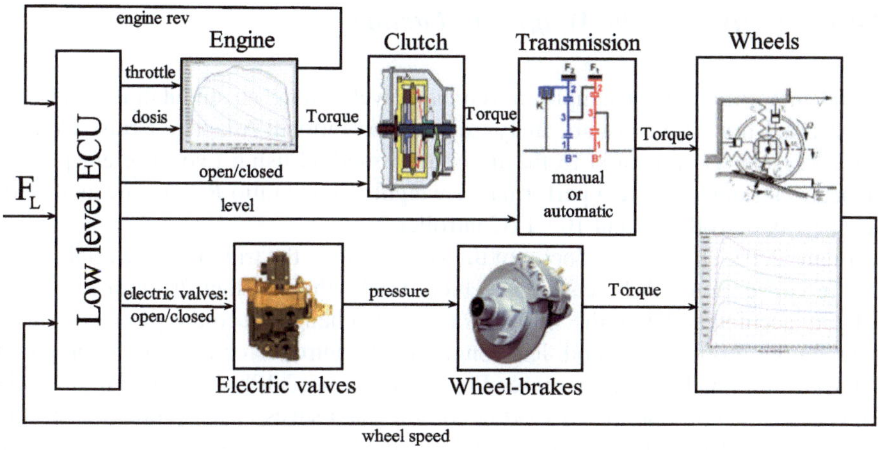

Fig. 2.9 Architecture of the low-level controller

low-level ECU might use more specific techniques that fit the specific nonlinear properties of the actuator. Figure 2.9 shows the architecture of the low-level ECU used in each vehicle.

The schematic structure of the system contains the next parts. The engine is controlled by the throttle, which could be a butterfly gate or a quantity of injected fuel. The engine-management system and the fuel-injection system have their own controllers, thus in the realization of the low-level controller only the torque-rev-load characteristics of the engine are necessary. In this case, the rev of the engine is measured, the required torque is computed from the longitudinal force of the high-level controller, thus the throttle is determined by an interpolation step using a look-up table. The position of the automatic transmission is determined by logic functions, thus it depends on the fuel consumption and the maximum rev of the engine. The pressures on the cylinders of the wheels increase during braking. At normal traveling, the ABS actuation is not necessary, in case of an emergency, the optimal driving is overwritten by the safety requirements. The necessary braking pressure for the required braking force is computed from the ratios of the hydraulic/pneumatic parts.

2.4 Simulation Examples

The purpose of the simulation examples is to illustrate the operation of the proposed method and present detailed analysis of the effects of the various parameters on the method. Finally, a 56-km-long highway with real data is analyzed.

2.4.1 Analysis of the Weighting Factors

In the first simulation example, a vehicle travels along an undulating road, see Fig. 2.10a. Three strategies are analyzed: the conventional Adaptive Cruise Control (ACC) system is compared to the introduced method using two different performance weighting. Different performance weights for the ratio $R_1 : R_2$ are analyzed, i.e., $1 : 1$ (Controller 1) and $10 : 1$ (Controller 2).

Figure 2.10c shows the velocity of the vehicle along the trip. It can be established that in conventional ACC case the vehicle tracks the constant 70 km/h reference velocity accurately, while the velocity of vehicle changes according to road conditions. In the proposed method, less longitudinal control force is enough compared to the conventional ACC system, see Fig. 2.10d. Figure 2.10b shows that the vehicle which considers the predicted road conditions also saves energy compared to the conventional ACC system. In order to qualify the controllers, the necessary absolute values of actuated control energies are added up in time ($\sum |F_l|$). The ACC controller which does not take into consideration the road conditions requires 12% more

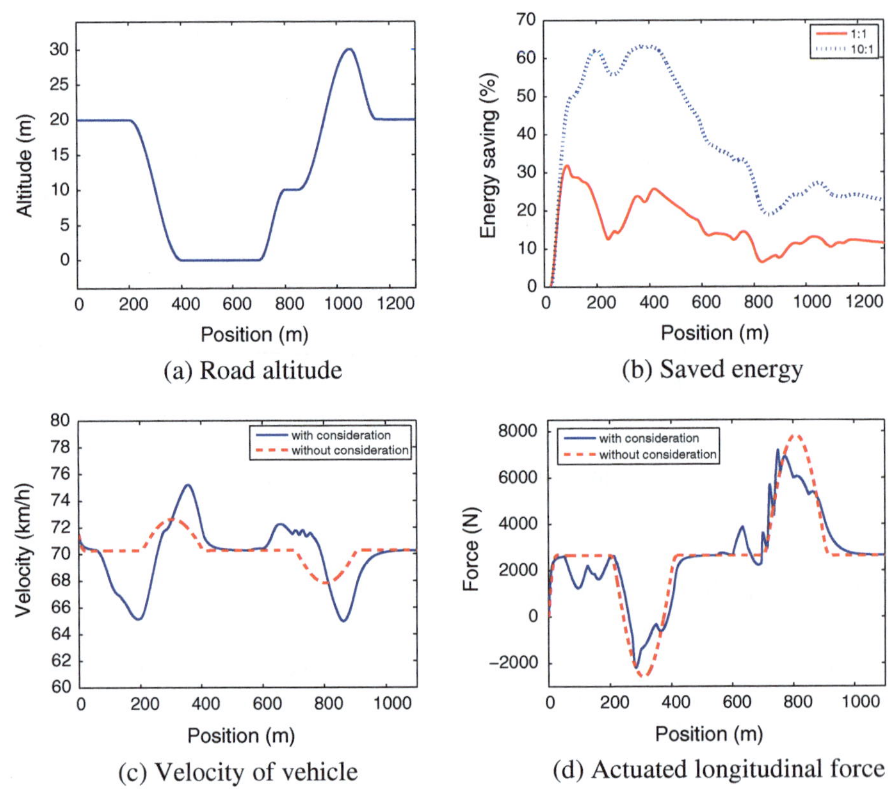

(a) Road altitude

(b) Saved energy

(c) Velocity of vehicle

(d) Actuated longitudinal force

Fig. 2.10 ACC systems on hilly road

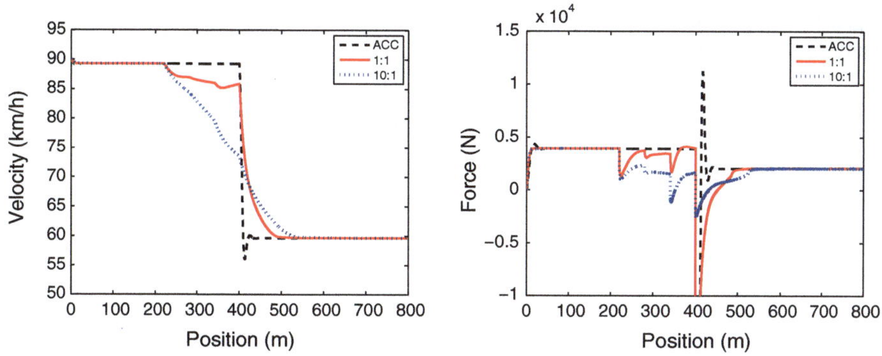

(a) Velocity (b) Actuated force

Fig. 2.11 ACC systems with a compulsory speed limit

control energy than Controller 1 and 23% more than Controller 2. The lower energy requirement of Controller 2 shows that the strategy seems to be more economical than in Controller 1 but saving energy causes lower velocity on the downhill slope, which is unfavorable in terms of cruising. For this reason, a balance between saving energy and a decrease in velocity must be obtained since more braking force is used.

The second simulation demonstrates a real traffic case, in which the vehicle is arriving at a section where a speed limit is imposed, thus the velocity must be reduced. The vehicle is traveling at 90 km/h and within 400 m its velocity must be reduced to 60 km/h. Figure 2.11 shows the simulation results.

If the controller is designed by using the road slope, the reference velocity changes continuously as Fig. 2.11a shows. The tendency of change depends on the proportion of performance weights, at $R_1 : R_2 = 10 : 1$ the change of velocity increases. At the same time, the velocity of vehicle in which the controller uses constant reference velocity (ACC) decreases rapidly. It can be stated that the knowledge of predicted road conditions control energy can be saved compared to ACC controller (see Fig. 2.11b). When $R_1 : R_2 = 1 : 1$ the saved energy is 19.7%, when $R_1 : R_2 = 10 : 1$ this value is 40%.

2.4.2 Impact the Various Parameters on the Adaptive Cruise Control

The following simulation example illustrates the effects of the various parameters on the proposed method. The effects of the performance weights, the number of sampling points and the vehicle mass are analyzed. In this simulation, the vehicles, travel on the same road, terrain characteristics of which are illustrated in Fig. 2.12. Figure 2.13 shows how the performance weights R_1 and R_2 influence the optimal

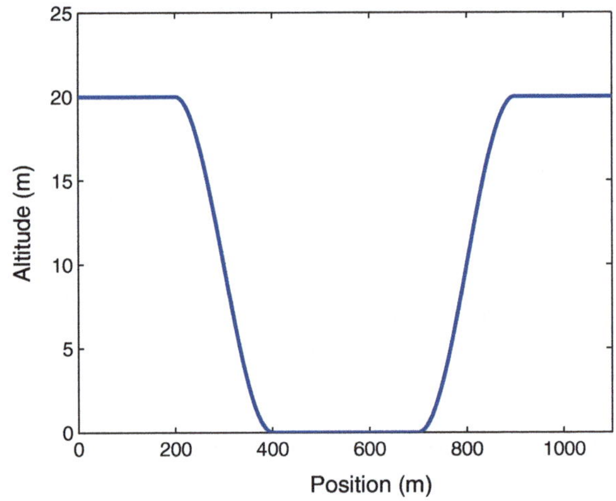

Fig. 2.12 Shape of the analyzed road section

velocity and the required longitudinal force. Different performance weights for the ratio R_1 and R_2 are analyzed. Figures show the operation of the proposed controller together with the operation of Controller 2. Since the performance weight R_1 focuses on the longitudinal force, it improves results by reducing force/energy requirements. At the same time, however, the controlled system pays less attention to the difference between the momentary velocity. Similarly, if R_2 increases the difference between the current and the reference velocities is less and the simulation results are closer to the results of Controller 2.

It is important to analyze the force/energy saving during the journey, see Fig. 2.13a. There are sections where energy saving is very significant, e.g., more than 30%. Energy savings in the downhill, uphill, and flat sections are different. In the downhill section, the vehicle has more possibility in the reduction of force/energy since here the longitudinal disturbances also decelerate the vehicle and there is less actuator requirement. In the uphill section, there is less possibility for the control system, since there are no additional forces which help the longitudinal actuator. The result shows that the driver is able to influence the different performances by tuning the performance weights and creating a balance between saving force/energy and traveling time.

In the analysis of the control system, different numbers of sampling points are selected, namely, 3, 4, and 5 equidistant points on a 150 m road. Figure 2.14 shows that this parameter influences the control system significantly. The number of sampling points determines the sampling length. When the sampling length is greater the vehicle is not able to take into consideration the effect of inclinations, since in the control design constant vehicle dynamics is assumed. Figure 2.14b shows that the velocity decreases significantly at the beginning of the downhill section and increases significantly at the uphill section if greater sampling length is selected. The results show that fewer sampling points result in saving more force/energy. It can also be seen that by increasing the sampling length, the simulation results differ more significantly

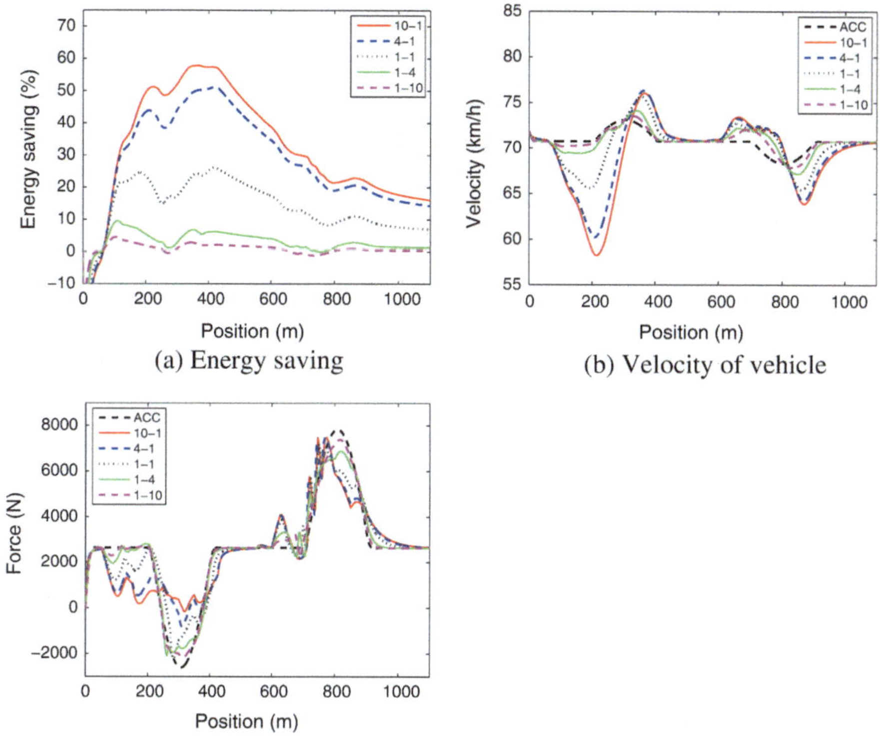

(a) Energy saving

(b) Velocity of vehicle

(c) Actuated force of vehicle

Fig. 2.13 Analysis of the performance weights based on different weighting strategies

from Controller 2. The analysis shows that there are relationships between the road conditions and sampling points (sampling length). For example, in the case of flat roads, it is enough to use fewer section points and greater sampling lengths because the slopes of the sections do not change abruptly. In the case of undulating roads, it is necessary to use more section points and smaller sampling length.

Figure 2.15 shows the mass dependence of velocity and longitudinal force. Four vehicle masses are analyzed: 1000, 3500, 8000, and 12000 kg. The other parameters are not changed during the simulations. Figure 2.15a shows the energy saving along the road section, which means that approximately 6–7% of energy can be saved by considering road conditions. The amount of energy saved increases slightly with the increase in the mass of the vehicle. Figure 2.15b shows that an increase in the vehicle mass makes velocity less abrupt. The computed longitudinal control forces of the vehicles are illustrated in Fig. 2.15c. It can be stated that increase in the vehicle mass improves the efficiency of the presented method, which means that at fixed performance weights R_1, R_2, the solution of optimization (see Sect. 2.2) is closer to the solution of performance criterion Optimization 1. It can be set back with the change of R_1, R_2 weights.

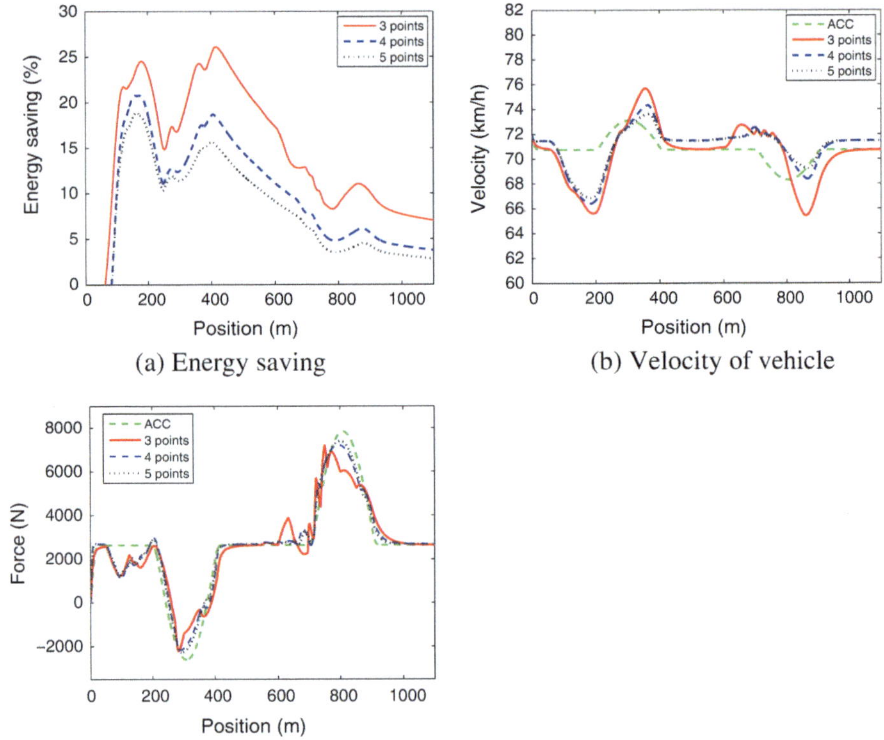

(a) Energy saving (b) Velocity of vehicle (c) Actuated force of vehicle

Fig. 2.14 Analysis of dependence of number of sampling points

2.4.3 Analysis of the Look-Ahead Method in a Motorway

In the last simulation, a transportational route with real data is analyzed. The terrain characteristics and geographical information are those of the M1 Hungarian highway between Tatabnya and Budapest in a 56-km-long section. In the simulation, a typical F-Class truck travels along the 56 km route. The mass of the 6–gear truck is 2023 kg and its engine power is 300 kW (402 hp). The regulated maximal velocity is 130 km/h, but the road section contains other speed limits (e.g., 80 km/h or 100 km/h), and the road section also contains hilly parts. Thus, it is an acceptable route for the analysis of road conditions, i.e., inclinations and speed limits. Publicly accessible up-to-date geographical/navigational databases and visualization programs, such as Google Earth and Google Maps are used for the experiment.

Figure 2.16a shows the altitude of the road along the way. It shows that the chosen motorway contains several uphill and downhill sections. In this example, two different controllers are compared. The first is the proposed controller, which considers the road conditions such as inclinations and speed limits and is illustrated by solid line in

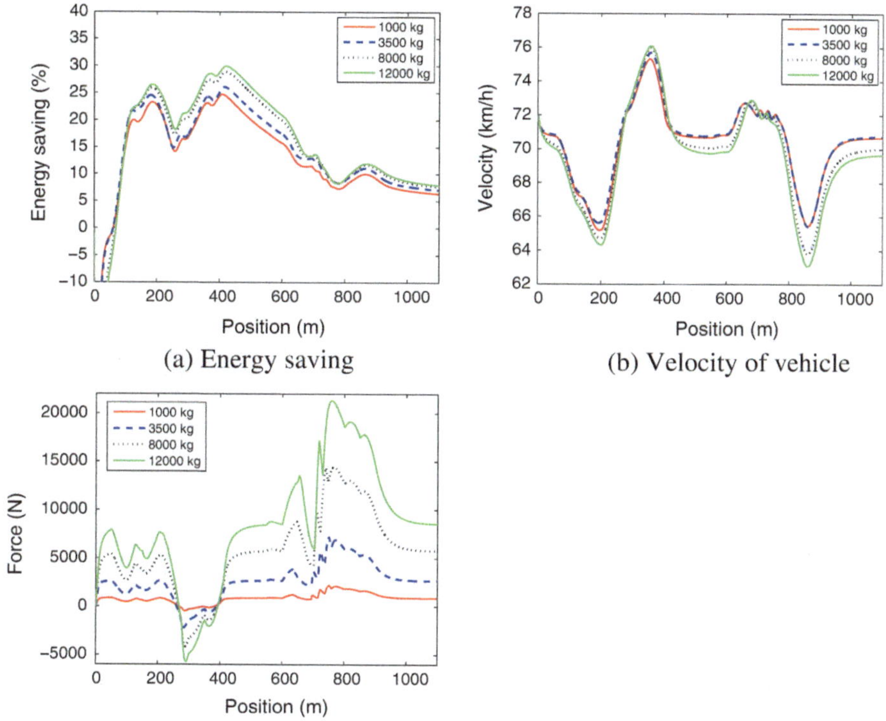

(a) Energy saving

(b) Velocity of vehicle

(c) Actuated force of vehicle

Fig. 2.15 Analysis of dependence of the mass

the figures, while the second controller is a conventional ACC system, which ignores this information and is illustrated by dashed line. Figure 2.16b shows the velocity of the vehicle with speed limits in both cases. The conventional ACC system tracks the predefined velocity speed limits as accurately as possible and the tracking error is minimal. In the proposed method, the velocity of the vehicle is determined by the speed limits and simultaneously it takes the road inclinations into consideration according to the optimal requirement. In the sections of road inclinations, the average relative difference between the actual velocity and the speed limit is 8% in the proposed method.

Figure 2.16c shows the required longitudinal force. The high-precision tracking of the predefined velocities in the conventional ACC system often requires extremely high forces with abrupt changes in the signals. Since the proposed method uses the road inclinations and speed limits in advance in the optimization method the truck is able to travel along the road with smaller actuation. The actual physical control inputs of the brake cylinder pressure and the throttle are illustrated in Fig. 2.16d and f, respectively. In the conventional ACC system, the revolution of the engine is within smaller bounds than in the proposed controller as Fig. 2.16e shows.

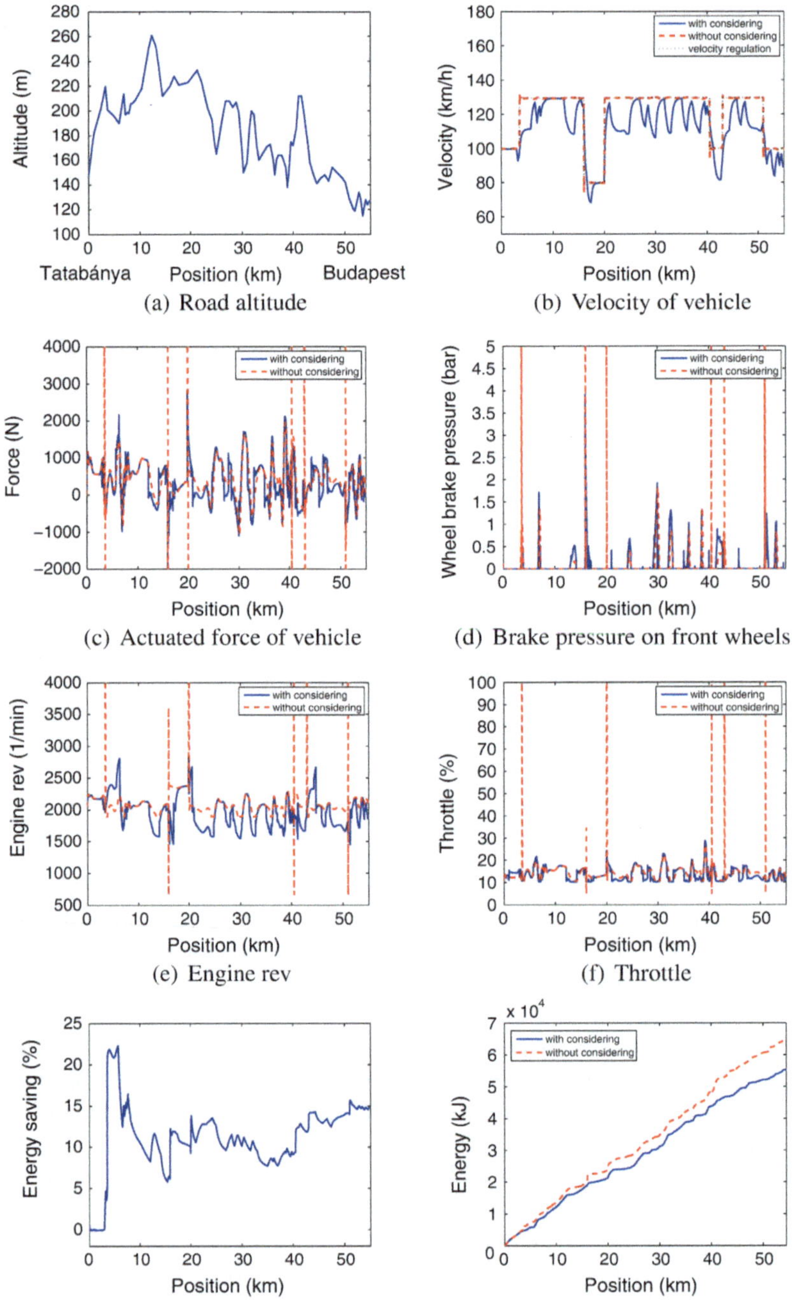

(a) Road altitude
(b) Velocity of vehicle
(c) Actuated force of vehicle
(d) Brake pressure on front wheels
(e) Engine rev
(f) Throttle
(g) Energy saving
(h) Total energy

Fig. 2.16 Real data motorway simulation

As a result of the road conditions, less energy is required during the journey in the proposed control method, see Figs. 2.16g and h. The proposed method requires smaller energy (47.2 MJ) than the conventional method (55.6 MJ), and the energy saving is 8.4 MJ, which is 15.1%. Approximately half of this value is realized by the driveline system (51.4%). The fuel consumption can also be calculated by using the following equation: $V = (F_{l1}\dot{\xi}_0)/(\eta L_h \rho_{fuel})$ where $\eta = 0.25$ is the efficiency of the driveline system, $L_h = 47.3$ MJ/kg is the heat of combustion and $\rho_{fuel} = 730$ kg/m^3 is the density of petrol. The fuel consumption of the conventional system is 6.44 l while that of the proposed method is 5.47 l, which results in 0.97 l reduction in fuel consumption in the analyzed 56 km length section. Since in this method, the velocity of the vehicle may be below the permitted maximum for the given section and accelerations/decelerations are carried out more slowly and gradually than in the conventional method the duration of the journey is expected to be longer. However, the difference in the duration is only 2 min.

Owing to the integration of vehicle actuators and road conditions, the simulation results show that the designed control reduces the energy required by the actuators and it slightly increases the traveling time. An important advantage of the method is that the procedure can be implemented in an Electronic Control Unit (ECU) in the practice of the vehicle industry.

2.4.4 Comparison with Dynamic Programming

The results of the proposed predictive cruise control are compared with those of a dynamic programming method through a simulation scenario. In the example, the vehicle must be driven along a 10-km-long segment of the Hungarian M1 highway between Budapest and Tatabnya, see Fig. 2.17. In the case of the dynamic programming algorithm, the following criteria must be guaranteed.

- The speed limitation is 130 km/h on the highway, and the maximum reduction of the vehicle speed is 20%, which results in 105 km/h minimum speed.
- The maximum speed of the vehicle is allowed to be maximum 1.5% higher, which results in 132 km/h.
- The vehicle must reach the maximum speed limit at the end of the highway section.

During the computation of the optimum speed profile, the 10-km-long section is divided into 200 equidistant sections with 50 m lengths. The goal of the optimization is to guarantee the minimization of the longitudinal control forces $F_{l,i}$ for the sum of the segments i

$$\min_{F_{l,1} \dots F_{l,200}} \sum_{i=1}^{200} |F_{l,i}| \tag{2.70}$$

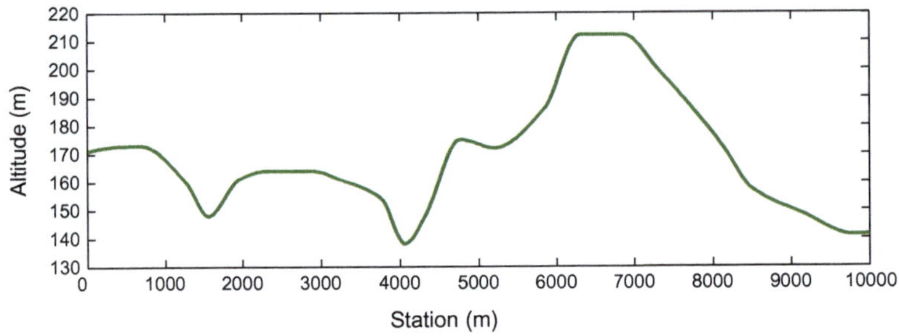

Fig. 2.17 Altitude of the highway segment

such that the previous criteria and the longitudinal dynamics of the vehicle (2.1) are guaranteed.

The results of dynamic programming, such as the speed profile and the control force are presented in Fig. 2.18. It can be seen that the resulting speed profile and force characteristics fit to the uphill/downhill sections of the highway (see Fig. 2.17), by which the acceleration and deceleration effects of the highway are utilized. The speed profile in Fig. 2.18a shows that in the section 3000 . . . 7000 km the speed profile is significantly influenced by the terrain characteristics. For example, at 4000

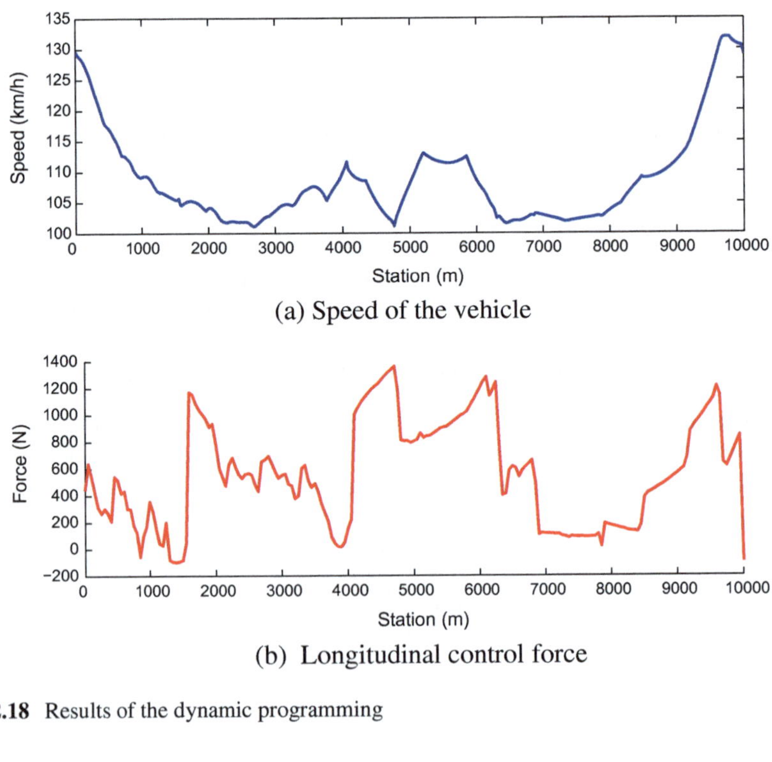

(a) Speed of the vehicle

(b) Longitudinal control force

Fig. 2.18 Results of the dynamic programming

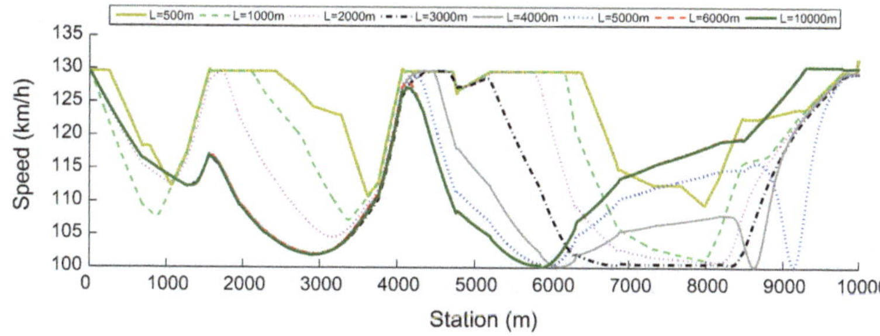

(a) Speed of the vehicle

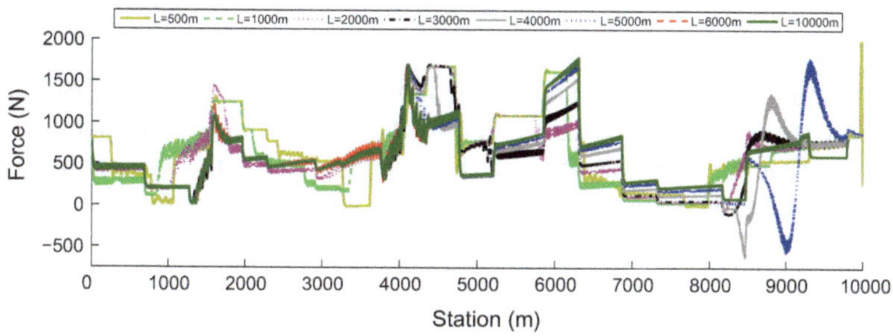

(b) Longitudinal control force

Fig. 2.19 Results of the proposed predictive cruise control

m there is valley, which results in the local maximum of the speed. Moreover, the longitudinal force is continuously reduced before the section point 4000 m, while the increased speed is advantageous for the uphill section of the exit of the valley.

Figure 2.19 illustrates the result of the proposed predictive cruise control with different look-ahead horizon lengths L. Due to the long highway section, the horizon significantly influences the speed profile of the vehicle. Although the optimization task of dynamic programming (2.70) with its constraints slightly differs from the proposed predictive cruise control strategy, it provides a good benchmark for the evaluation of the method. In the case of a short horizon (500 . . . 2000 m), the speeds of the vehicle often reach the speed limit 130 km/h, while in the long-horizon scenarios (5000 . . . 10000 m), the variation of the terrain characteristics has higher impact on the speed profile. Moreover, with increasing horizons, the speed profiles approximate the result of dynamic programming. In the control forces, a similar tendency can be found, see Fig. 2.19b. Thus, the proposed predictive cruise control method provides an appropriate approximation of dynamic programming.

The energy consumption values in the various methods are illustrated in Fig. 2.20. It shows that the energy consumption at the predictive cruise control is approximately unchanged at $L > 3000$ m. Thus, it is sufficient to consider information from 3000 . . . 4000 m ahead of the vehicle. Moreover, the comparison of the global dynamic programming and the predictive cruise control demonstrate that the pro-

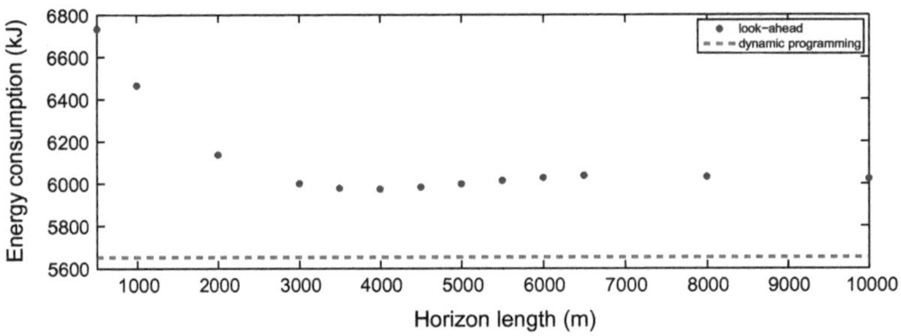

Fig. 2.20 Energy consumption depending on the horizon length

posed method provides a good approximation in the results of the energy consumption. Predictive cruise control results in only 7% energy increase during the simulation at $L = 3000$ m. However, the huge advantage of predictive cruise control is the reduction of the computation effort regarding dynamic programming.

2.4.5 *Stability Analysis*

The reference signal generator of the control design is illustrated through a simulation scenario with a real highway topographic data. The highway section is a part of the French highway *A*36 in Alsace, between Mulhouse and Belfort, in which the speed regulations vary, see Fig. 2.21a and b. The nominal weight of the vehicle is 3500 kg. The predicted road horizon in the look-ahead control is 500 m long, which is divided into 10 subsections. The aim of the simulation is to illustrate both the tracking performance and the stability of the closed-loop system.

The current speed of the vehicle and the generated reference speed are shown in Fig. 2.21b. In order to minimize the fuel consumption and time requirements, the design of the optimal speed profile is based on the uphill and downhill sections and the speed limits. The speed signals show that the controller guarantees an adequate tracking performance. The control input F_{l1} guarantees robust performances as Fig. 2.21c shows.

The results of the identification concerning the reference speed generator is illustrated in Fig. 2.22. The parameters of the transfer functions

$$G_{genA} = \frac{B_{genA,1} + B_{genA,2}q^{-1}}{1 + A_{gen,1}q^{-1} + A_{gen,2}q^{-2}},$$

$$G_{genB} = \frac{B_{genB,1} + B_{genB,2}q^{-1}}{1 + A_{gen,1}q^{-1} + A_{gen,2}q^{-2}}.$$

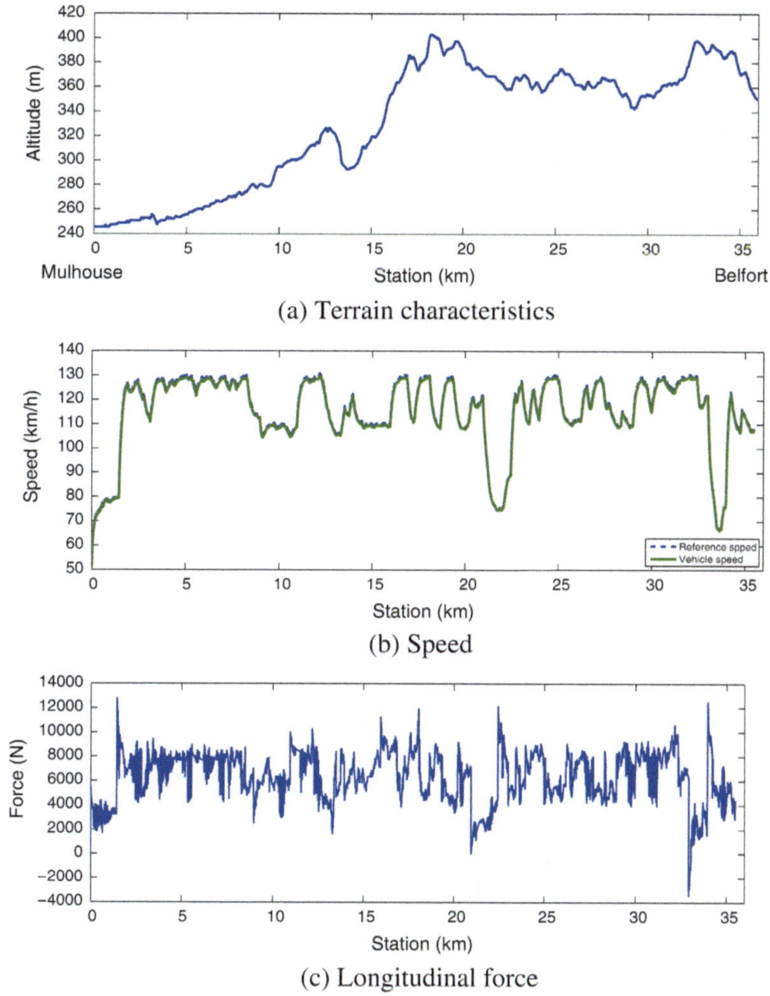

Fig. 2.21 Comparison of performances

are illustrated in Fig. 2.22a and b, respectively. The parameters are almost constant and their significant changes are caused by the sharp changes in the altitude or the speed limits.

The identification of G_{gen} is used to verify the stability of the closed-loop system. The computed infinite norm of the closed-loop system is illustrated in Fig. 2.23. Since the infinite norm is under the critical value 1, robust stability of the entire system is guaranteed.

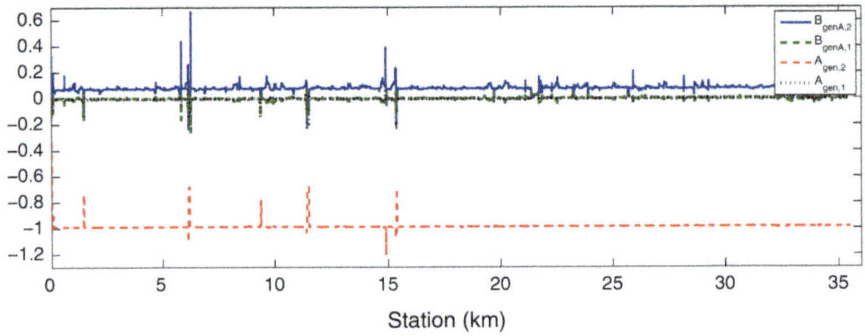

(a) Parameters of G_{genA}

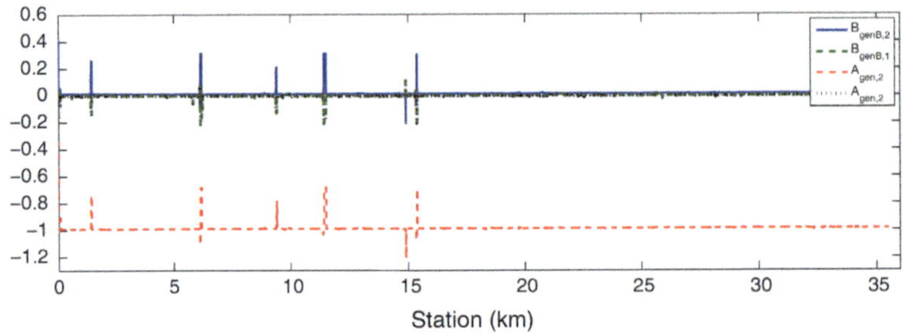

(b) Parameters of G_{genB}

Fig. 2.22 Identified parameters of G_{gen}

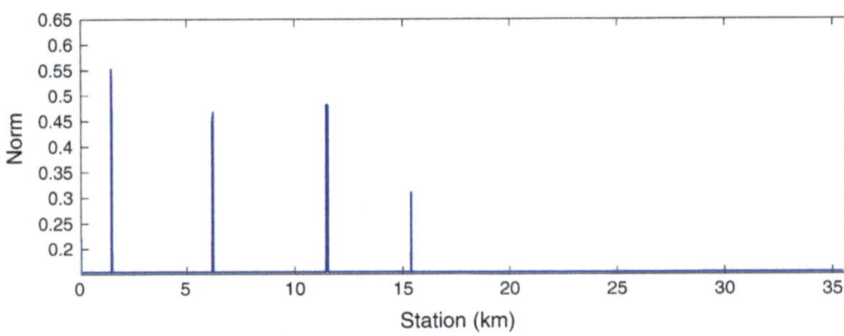

Fig. 2.23 Infinite norm of the closed-loop system

References

Bae HS, Ruy J, Gerdes J (2001) Road grade and vehicle parameter estimation for longitudinal control using GPS. In: 4th IEEE Conference on Intelligent Transportation Systems, vol 1, pp 1–6

Bokor J, Balas G (2005) Linear parameter varying systems: a geometric theory and applications. In: 16th IFAC world congress, Prague, vol 1, pp 1–12

Ebnre A, Hermann R (2001) A self-organized radio network for automotive applications. In: 8th World congress on intelligent transportation systems, Sydney, Australia

Faris WF, Rakha HA, Kafafy RI, Idres M, Elmoselhy S (2011) Vehicle fuel consumption and emission modelling: an in-depth literature review. Int J Veh Syst Model Test 6(3):318–395

Festag A, Hessler A, Baldessari R, Le L, Zhang W, Westhoff D (2008) Vehicle-to-vehicle and road-side sensor communication for enhanced road safety. In: 15th World congress on intelligent transport systems

Hahn J, Rajamani R, You S, Lee K (2004) Real-time identification of road-bank angle using differential GPS. IEEE Trans Control Syst Technol 12:589–599

Hellström E, Ivarsson M, Åslund J, Nielsen L (2009) Look-ahead control for heavy trucks to minimize trip time and fuel consumption. Control Eng Pract 17(2):245–254

Hellström E, Åslund J, Nielsen L (2010) Horizon length and fuel equivalents for fuel-optimal look-ahead control. Adv Automot Control 43(7): 360–365

Ivarsson M, Åslund J, Nielsen L (2009) Look ahead control - consequences of a non-linear fuel map on truck fuel consumption. Proc Inst Mech Eng Part D, J Automob Eng 223:1223–1238

Kesting A, Treiber M, Schnhof M, Helbing D (2007) Extending adaptive cruise control to adaptive driving strategies. Transp Res Rec 2000:16–24

Kiencke U, Nielsen L (2000) Automotive control systems for engine, driveline and vehicle. Springer, Heidelberg

Kolmanovsky I, Filev D (2009) Stochastic optimal control of systems with soft constraints and opportunities for automotive applications. In: IEEE conference on control applications

Kolmanovsky I, Filev D (2010) Terrain and traffic optimized vehicle speed control. In: IFAC advances in automotive control conference

Labayrade R, Aubert D, Tarel J (2002) Real time obstacle detection in stereovision on non flat road geometry through "v-disparity" representation. Intell Veh Symp IEEE 2:646–651

Lingman P, Schmidtbauer B (2002) Road slope and vehicle mass estimation using Kalman filtering. Veh Syst Dyn Suppl 37:12–23

Ljung L (1999) System identification: theory for the user. Prentice Hall

Marquardt D (1963) An algorithm for least-squares estimation of nonlinear parameters. J Soc Ind Appl Math 11(2):431–441

Németh B, Gáspár P (2010) Considering predicted road conditions in vehicle control design using $\mathscr{H}_\infty$ method. In: IFAC advances in automotive control conference

Nouveliere L, Braci M, Menhour L, Luu H (2008) Fuel consumption optimization for a city bus. In: UKACC control conference

Nuevo J, Parra I, Sjoberg J, Bergasa L (2010) Estimating surrounding vehicles' pose using computer vision. In: 13th IEEE conference on intelligent transportation systems, pp 1863–1868

Pacejka HB (2004) Tyre and vehicle dynamics. Elsevier Butterworth-Heinemann, Oxford

Packard A, Balas G (1997) Theory and application of linear parameter varying control techniques. In: American control conference, workshop I, Albuquerque, New Mexico

Passenberg B, Kock P, Stursberg O (2009) Combined time and fuel optimal driving of trucks based on a hybrid model. In: European control conference, Budapest

Rakha H, El-Shawarby I, Arafeh M, Dion F (2006) Estimating path travel-time reliability. In: IEEE intelligent transportation systems conference, Toronto, Canada, pp 236–241

Rakha H, Van Aerde M, Case E, Ugge A (1989) Evaluating the benefits and interactions of route guidance and traffic control strategies using simulation. In: Vehicle navigation and information systems conference, 1989 Conference Record, pp 296–303

Sahlholm P, Johansson K (2009) Road grade estimation for look-ahead vehicle control using multiple measurement runs. Control Eng Pract (in press)

Swaroop D, Hedrick J (1996) String stability of interconnected systems. IEEE Trans Autom Control 41:349–357

Trachtler A (2004) Integrated vehicle dynamics control using active brake, steering and suspension systems. Int J Veh Des 36:1–12

Wu F, Yang X, Packard A, Becker G (1996) Induced L_2 norm controller for LPV systems with bounded parameter variation rates. J Robust Nonlinear Control 6:983–988

Zhou K, Doyle J, Glover K (1996) Robust and optimal control. Prentice Hall

Chapter 3
Design of Predictive Cruise Control Using Road and Traffic Information

Introduction and Motivation

The development of an energy-efficient operation strategy for road vehicles has been in the focus of research and development centers, suppliers, and manufacturers. The purpose of the strategy is to design the speed of road vehicles, in which several factors are taken into consideration such as control energy requirement, fuel consumption, road slopes, speed limits, emissions, and traveling time. The multi-objective optimization criteria are handled by the so-called look-ahead control methods. In the method, local traffic information is also taken into consideration in the control design. Thus, the look-ahead control can be considered as an extension of adaptive cruise control with road and traffic information.

Several papers have been published in these topics. The speed design for road vehicles based on road inclinations, speed limits, a preceding vehicle in the lane, and traveling time was proposed by Németh and Gáspár (2013b, 2014). A method for the design of a decision algorithm for lane-change maneuvers based on the consideration of surrounding vehicles was elaborated in Gáspár and Németh (2015). An eco-cruise control strategy in which the multi-criteria optimization between journey time and fuel consumption was converted into a constrained fuel optimization task was proposed by Saerens et al. (2013). A predictive cruise control with upcoming traffic signal information to improve fuel economy and reduce traveling time was proposed by Asadi and Vahidi (2011). The algorithm used radar and traffic signals to optimize traveling speed. Estimation methods of surrounding vehicle motion prediction for safety critical traffic applications were proposed in Barth and Franke (2008), Broadhurst et al. (2004). In the case of hybrid electric vehicles, road prediction is important to optimize battery recovery. The predictive reference signal generator method to maximize recuperated energy using the topographic profile of the future road segments and the corresponding average traveling speeds was proposed by Ambuhl and Guzzella (2009). The shape of the speed profile at a road segment

P. Gáspár and B. Németh, *Predictive Cruise Control for Road Vehicles Using Road and Traffic Information*, Advances in Industrial Control,
https://doi.org/10.1007/978-3-030-04116-8_3

was predefined, and its parameters were determined during a nonlinear constrained optimization process in van Keulen et al. (2010).

Although the efficiency of terrain preview consideration in cruise control has been illustrated in several publications, a smaller emphasis has been placed on traffic flow impair caused by speed reduction. The driver of the look-ahead vehicle is able to create a balance between longitudinal energy saving and journey time according to his own priorities. However, other drivers on the road have different priorities, which can lead to conflict, e.g., fast vehicles are held up by vehicles traveling in a fuel-efficient fashion. The goal of the research is to design an optimal look-ahead control strategy which is able to adapt to the motion of the surrounding vehicles. The combination of the look-ahead concept and the congestion problem leads to a complex multi-criteria optimization task.

The chapter proposes the consideration of road and traffic factors in the look-ahead control in order to achieve an energy-efficient cruise control strategy for speed design. The multi-objective optimization results in weighting factors which are incorporated into the consideration of surrounding vehicles. In this way, a balance between the designed speed and the flow of the traffic can be guaranteed.

3.1 Handling the Preceding Vehicle in the Speed Design

Since the vehicle may catch up with a preceding vehicle, it is necessary to consider speed of the latter v_{lead}. The estimation of the safe stopping distance may be conservative in a normal traffic situation, where the preceding vehicle also brakes; therefore, the distance between the vehicles may be reduced.

The safe stopping distance between the vehicles is determined according to the 91/422/EEC, 71/320/EEC EU directives (in case of M_1 vehicle category, the velocity in km/h):

$$d_{st} = 0.1\dot{\xi}_0 + \dot{\xi}_0^2/150, \tag{3.1}$$

in which $\dot{\xi}$ is the current speed. It is also necessary to consider that without the preceding vehicle the consideration of safe stopping distance is neither possible nor necessary.

The consideration of the preceding vehicle is determined by W, which is selected in the following structure:

$$W = \begin{cases} 1 & \text{if} \quad d < d_{st} \\ (\alpha d_{st} - d)/((\alpha - 1)d_{st}) & \text{if } d_{st} \leq d \leq \alpha \cdot d_{safe} \\ 0 & \text{if} \quad d > \alpha \cdot d_{st} \end{cases} \tag{3.2}$$

where α is a design parameter, e.g., $\alpha = 0.5$ means that the speed of the vehicle must be reduced when the distance is less than $1.5 \cdot d_{st}$. The weight is illustrated in Fig. 3.1.

The velocity of the vehicle is calculated by using the optimization procedure. It is not modified until the distance from the preceding vehicle is greater than the

Fig. 3.1 Selection of weight W

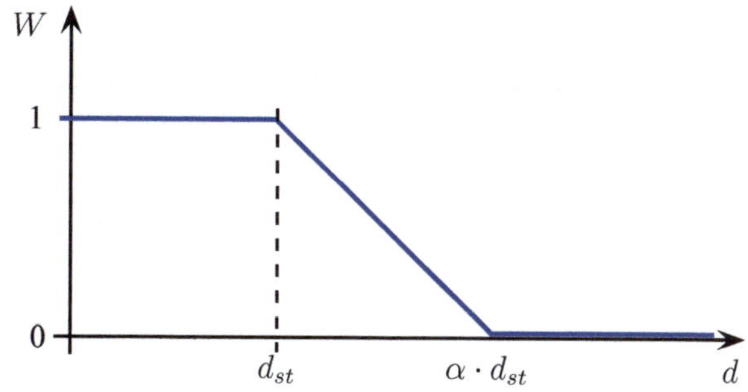

predefined value αd_{st}. However, when the distance is reduced, i.e., $d \leq \alpha d_{st}$, the velocity must be modified by applying the weighting factor W in the following way:

$$\lambda = W \cdot v_{lead}. \tag{3.3}$$

Consequently, the distance between the vehicles decreases to the safety value. This velocity must be applied, but the velocity is calculated by using the optimization procedure. When the preceding vehicle accelerates and exceeds the calculated velocity, then the optimal speed can be applied again.

Because of safety reasons, factor W has priority over the weights Q and γ_i; thus, in the weighting strategy instead of (2.19), weights should sum up to

$$\gamma_1 + \gamma_2 + ... + \gamma_n + Q = 1 - W. \tag{3.4}$$

and the new weights must be reduced in the following way:

$$Q_n = Q(1 - \mathscr{W}), \quad \gamma_{n,i} = \gamma_i(1 - W), \; i \in [1, n], \tag{3.5}$$

where Q and γ_i are the original weight factors. The designed speed of the vehicle does not exceed the speed defined by (3.3).

3.2 Considering the Motion of the Follower Vehicle in the Speed Design

3.2.1 Calculation of Safe Distance

Since the vehicle preferring energy saving travels in traffic, it may be in conflict with other vehicles preferring cruising at the speed limit. Preferring weight R_1 leads to a nonoptimal motion for the traffic. A balance between the energy-efficient speed and the flow of the local traffic is proposed by the modification of the value R_1. The

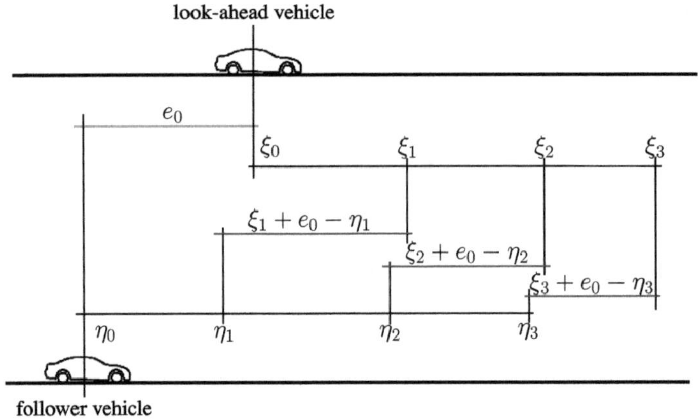

Fig. 3.2 Prediction of the vehicle motion

motion of follower vehicles is considered in three steps. First, the speed profile of the predictive control is predicted on the horizon, see Fig. 3.2.

Step 1: The speed prediction of the vehicle using look-ahead control is based on (2.26). Based on (2.27), the expression of ϑ can be rewritten as

$$\begin{aligned} \vartheta &= v_{ref,0}^2 - R_1(1-\bar{Q})v_{ref,0}^2 + R_1\sum_{i=1}^{n}\bar{\gamma}_i v_{ref,i}^2 + R_1\left(\frac{2}{m}\sum_{i=1}^{n} s_i F_{di,r}\sum_{j=i}^{n}\bar{\gamma}_j\right) \\ &= v_{ref,0}^2(1-R_1) + R_1\bar{\vartheta}, \end{aligned} \tag{3.6}$$

where $\bar{\vartheta}$ contains the value of ϑ calculated with energy-efficient prediction weights $\bar{Q}, \bar{\gamma}_i$. From (2.32), the calculation of the reference speed λ is based on the predicted road information. Through Q and ϑ performance weight, R_1 plays an important role in the calculation of the reference speed. Moreover, the predicted values of the prediction weights γ_i also depend on R_1, see (2.47)–(2.49). The square of the reference speed is calculated in the following form:

$$\begin{aligned} \lambda^2 &= v_{ref,0}^2(1-R_1) + R_1\bar{\vartheta} - 2s_1R_1(1-\bar{Q})(\ddot{\xi}_0 + g\sin\alpha) \\ &= v_{ref,0}^2(1-R_1) + R_1\bar{\lambda}^2. \end{aligned} \tag{3.7}$$

Thus,

$$\lambda = \sqrt{v_{ref,0}^2(1-R_1) + R_1\bar{\lambda}^2}, \tag{3.8}$$

where $\bar{\lambda}$ contains the value of λ calculated with energy-efficient prediction weights $\bar{Q}, \bar{\gamma}_i$. From (2.15) and (3.7), the predicted speed of the vehicle at section point n is

$$\begin{aligned}\dot{\xi}_n^2 =& v_{ref,0}^2(1 - R_1) + R_1\bar{\lambda}^2 + \frac{2}{m}s_1 F_{l1} - \frac{2}{m}s_1 F_{d1,o} - \frac{2}{m}\sum_{i=1}^{n} s_i F_{di,r} \\ =& R_1\mathcal{N}_1 + \mathcal{N}_2. \end{aligned} \tag{3.9}$$

According to (3.9), the predicted speed $\dot{\xi}_n$ at point n is independent of $v_{ref,n}$. However, when $R_1 = 0$, the predicted speed at point n must be $v_{ref,n}$. In order to meet this requirement, the predicted speed must be modified by using the reference speed and the weighting factor in the following way:

$$\dot{\xi}_n^2 = (R_1\mathcal{N}_1 + \mathcal{N}_2)R_1 + (1 - R_1)v_{ref,n}^2. \tag{3.10}$$

In the formula, $\mathcal{N}_1$ is independent of the number of section points ahead, while $\mathcal{N}_2$ contains the road grade information of each section. The advantage of this formula is that the reference speed is built into the predicted speed, and thus the numerical procedure is more reliable.

Step 2: In the second step, the motion of the follower vehicle is predicted. In the estimation of the follower vehicle, several assumptions are made. First, the controlled vehicle has information about the speed and acceleration of the follower vehicle ($\dot{\eta}_0$, $\ddot{\eta}_0$) and the momentary distance between the vehicles e_0. Second, the follower vehicle accelerates evenly until it reaches the speed limit, i.e., $i < j$. Then, it does not accelerate further; thus, in the oncoming sections, the predicted speeds of the vehicle are $v_{ref,j}, \ldots, v_{ref,n}$, i.e., $i \geq j$.

Based on the information ($\dot{\eta}_0$, $\ddot{\eta}_0$, e_0), the motion of the vehicle must be calculated in every section in which the traveling time is $\Delta t_i, i = \{1 \ldots n\}$. During acceleration, the displacements of the follower vehicle are

$$\eta_k = \frac{\ddot{\eta}_0}{2}\left(\sum_{i=1}^{k}\Delta t_i\right)^2 + \dot{\eta}_0\sum_{i=1}^{k}(\Delta t_i), \quad k \in [1, j-1] \tag{3.11}$$

in which traveling time in a section is Δt_i. When the follower vehicle reaches the speed limit, it does not accelerate further. Moreover, its speeds do not exceed the predefined reference speeds $v_{ref,j}, \ldots, v_{ref,n}$. The displacements of the vehicle are

$$\eta_l = \eta_{j-1} + \sum_{i=j}^{l}(v_{ref,i}\Delta t_i), \quad l \in [j, n]. \tag{3.12}$$

Step 3: Finally, in the third step, the safe distance is calculated. Now, the safe distance between the controlled vehicle and the follower one must be guaranteed. The safety distance d_{safe} is assumed to be predefined.

The controlled vehicle intends to use the energy-efficient predicted cruise control, while the follower vehicle aims to keep the speed limit. Thus, the look-ahead control strategy is modified in such a way that the motion of the follower vehicle is taken into consideration. A possible method is to modify performance weight R_1 during the journey and create a balance between the designed speed and the required speed of the follower vehicle. The aim of this section is to develop a method for the redesign of weight R_1.

The criterion of the safety distance is based on the motion of the vehicles. During the journey in every section, the distance between the two vehicles must be guaranteed by the following inequalities:

$$\xi_i + e_0 - \eta_i \geq d_{safe}, \quad i \in \{1, 2, .., n\}, \tag{3.13}$$

where ξ_i is the predicted displacement of the controlled vehicle, e_0 is the momentary distance between the vehicles ($t = 0$), and η_i is the predicted displacement of the follower vehicle. It is necessary to find the maximum of performance weight R_1, which satisfies the inequality constraints (3.13). Note that an increase in R_1 induces longer journey time. Therefore, R_1 can be limited by the driver using a predefined bound $R_{1,max}$.

3.2.2 *Optimization for Safe Cruising*

The optimization criterion for safe cruising is formulated as follows:

$$\max_{[0, R_{1,max}]} R_1, \tag{3.14}$$

such that the following conditions are satisfied:

$$\sum_{i=1}^{j} \xi_i + e_0 - \eta_j - d_{safe} \geq 0, \quad j \in \{1, ..., n\}. \tag{3.15}$$

The result of the optimization $R_{1,opt}$ is used in the calculation of the prediction weights Q and γ_i. Based on the prediction weights, the reference speed of the controlled vehicle λ is computed. The optimization procedure (3.14) is performed in each step, and thus performance weight R_1 is rewritten continuously according to the current local traffic information.

In practice, the solution of the optimization procedure (3.14) requires a great deal of computation effort. However, in solution (3.14), the result of the previous computation step $R_{1,old}$ can be applied as initial value. If the new solution $R_{1,new}$ is searched for in the interval

$$[\max(R_{1,old} - \alpha, 0), \min(R_{1,old} + \alpha, R_{1,max})],$$

the computation time is reduced significantly. Note that $R_{1,max}$ is set by the driver within the interval $R_{1,max} = [0; 1]$.

3.3 Lane Change in the Look-Ahead Control Concept

In this section, the decision algorithm of the lane change is proposed. During the lane change, safe operation must be guaranteed and the conflicts between vehicles and tailbacks must be prevented. First of all, the safe distance from both the preceding vehicle and the follower one in the new lane must be ensured. Consequently, the safety distance must be examined before the lane change is carried out. Second, the conflicts between the vehicle preferring energy saving and the follower vehicle preferring cruising at the speed limit must be avoided. Thus, handling the preceding vehicle and considering the motion of the follower vehicle must be incorporated into the decision method.

Essentially, the lane change is preferred if the vehicle using energy-efficient speed control is not able to keep the designed speed in the current lane and at the same time the designed speed can be ensured and the maneuver is safe in the adjacent lane. The speed profile is influenced by two weighting factors in the look-ahead control. Weight R_1 is used to consider the follower vehicle, while weight W is used to handle the preceding vehicle. Besides the calculation of the optimal speed, during the journey the acceptable speed is computed for both the current lane and adjacent lane. Since the weighting factors R_1 and W influence the acceptable speed, they are also computed and considered. In the decision method of the lane change, these two factors must be analyzed.

Scenario 1: The controlled vehicle catches up with a preceding vehicle. Since W is increased, the vehicle adapts to the velocity of the preceding vehicle. Since the preceding vehicle is traveling slower, the current speed may differ significantly from the designed velocity, and thus the lane change must be analyzed. W_{new} must be computed for the adjacent lane.

Scenario 2: The follower vehicle catches up with the controlled vehicle. Since R_1 is reduced, the vehicle adapts to the velocity of the follower vehicle. Since the follower vehicle is traveling faster, the current speed may differ significantly from the designed velocity, and thus the lane change must be analyzed. $R_{1,new}$ must be computed for the adjacent lane.

Two basic inequalities must be checked. W_{new} and $R_{1,new}$ are computed for the adjacent lane and they are compared to the current W and R_1. The lane change must be carried out if the following two inequalities persist. W_{new} is significantly smaller than W, which means that in the adjacent lane there is not a preceding vehicle and at the same time $R_{1,new}$ is larger than R_1, which means that the follower vehicles are not hindered and the energy-efficient velocity can be ensured. Moreover, the safety distances from the preceding vehicle and the follower one in the new lane can also be guaranteed. d_{pr} and d_{fo} are the distances from the preceding vehicle and the follower one in the new lane, respectively.

Summarizing the above thoughts, the following logic-based decision method can be formed:

$$\text{If } W - W_{new} > \varepsilon_W \text{ and } R_{1,new} - R_1 > \varepsilon_R, \tag{3.16}$$

and, moreover,

$$\text{if } d_{pr} \geq d_{safe} \text{ and } d_{fo} \geq d_{safe}, \tag{3.17}$$

where ε_W and ε_R are predefined value, then the lane change should be performed.

3.4 Simulation Results

3.4.1 Handling the Preceding Vehicle

In the examples, the controller in which the velocity profile is calculated by taking the road inclinations and speed limits into consideration (Controller 1) is compared to the controller in which the reference velocity is selected constant and the road information is not considered, i.e., it is a conventional adaptive cruise control (Controller 2).

Figure 3.3a shows the coordinates of the undulating road. It contains several uphill and downhill sections; therefore, it is suitable for the analysis of the proposed method. The reference velocity is usually $v_{ref} = 80$ km/h, but there is a speed limit on the road, i.e., between 1100 and 1300 m, and the reference velocity is $v_{ref} = 50$ km/h. Figure 3.3b shows the velocity of the vehicle. The dashed line illustrates that Controller 2 tracks the predefined constant velocities with an acceptable error and at the downhill and uphill sections the tracking error slightly increases due to the effect of road inclinations. At the speed limit signs, the velocity of the vehicle decreases and increases rapidly. At the same time, the velocity in the front of the downhill section is decreased by Controller 1, since the controller reduces the required longitudinal force, see Fig. 3.3c. It saves energy because the decelerating effects of road disturbances are exploited when the road conditions are considered. The downhill section has an accelerating effect; therefore, the velocity loss at the beginning of the downhill section can be compensated for. In the case of Controller 2, the braking system is also used to prevent the velocity increasing, while in the proposed method less braking operation is needed. In this case, Controller 1 considers that after the downhill section there is a flat road section and the deceleration effect of the road disturbances can compensate for the increase of velocity. At the downhill section, the consideration of the inclinations cannot be exploited by the conventional method, because the disturbances do not reduce the longitudinal force. In this section, Controller 2 also tracks the constant velocity. By considering the road conditions, the vehicle accelerates before the uphill section; therefore, it can reach a velocity at which the vehicle can travel along the uphill section with less longitudinal force. Figure 3.3c shows that Controller 1 requires less longitudinal force, and its maximal values are approximately half of the forces of Controller 2. At point 1100 m, the velocity of the vehicle is decreased before the speed limit sign and the velocity changes.

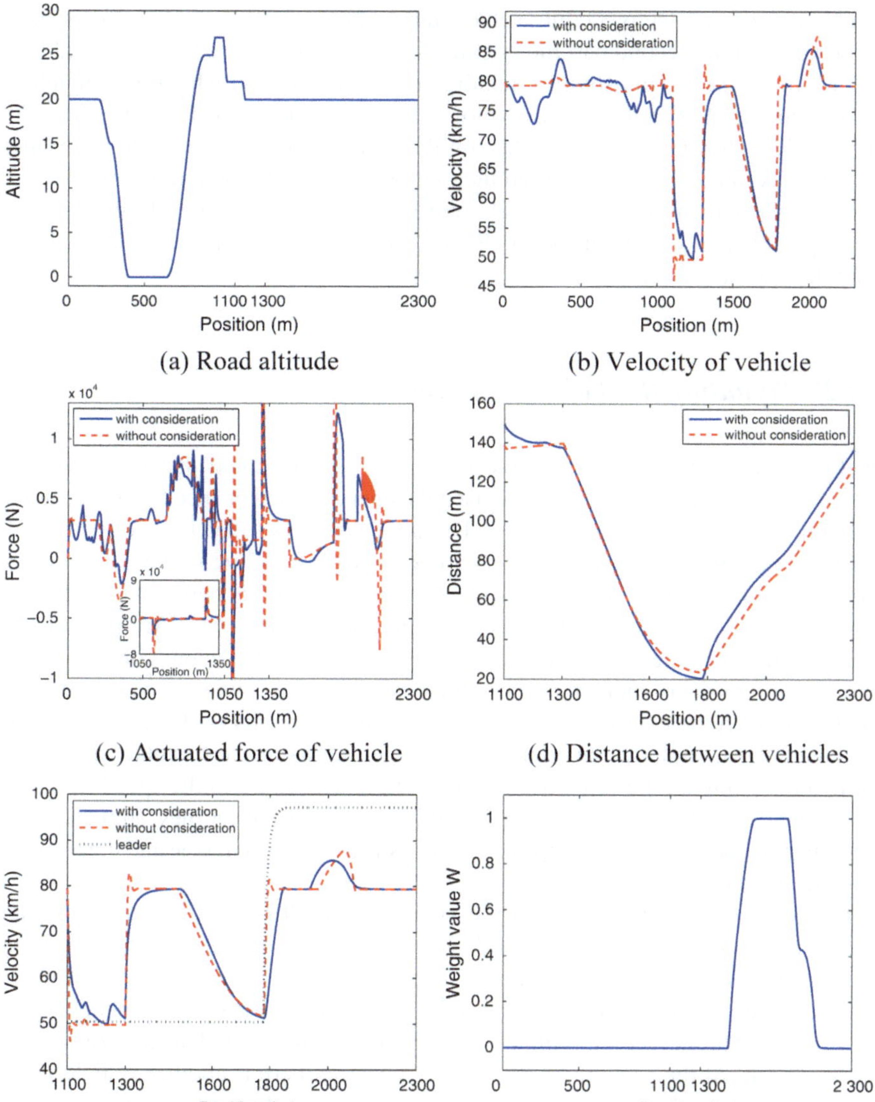

(a) Road altitude
(b) Velocity of vehicle
(c) Actuated force of vehicle
(d) Distance between vehicles
(e) Velocity of preceding vehicle
(f) Value of W

Fig. 3.3 Cruise control systems on hilly road

The simulation example also shows the effect of a preceding vehicle. The preceding vehicle travels only at 50 km/h instead of 80 km/h. Consequently, the distance between the vehicles decreases, see Fig. 3.3d. Later, the preceding vehicle accelerates and exceeds 80 km/h, see Fig. 3.3e. This simulation shows that the proposed cruise control method is able to adapt to the preceding vehicle, because the tracking vehicle approaches the preceding vehicle taking the braking distance into consideration, while in the second part the tracking vehicle avoids exceeding the speed limit and it falls behind. This solution requires radar information which is available in a conventional ACC vehicle. This velocity control is achieved by using the value of W as it is shown in Fig. 3.3f. The simulation example shows that the controller, which takes into consideration road inclination and speed limits, requires 14% less control energy than the conventional controller and the maximum difference from the original velocity is about 7 km/h. Besides, the designed control system is able to adapt to external circumstances.

3.4.2 Handling the Follower Vehicle

In this section, the efficiency of the method is demonstrated during a relatively complex traffic scenario. The vehicle used the energy-efficient speed control catches up with a slower preceding vehicle. The analysis shows that the slower vehicle must be overtaken. However, there is another vehicle traveling in the adjacent lane within the safe distance, see Fig. 3.4a. Since there is a conflict between vehicles, the controlled vehicle must wait until the adjacent lane becomes free. At that moment, the preceding vehicle is overtaken by engaging the energy-saving control. At the same time, however, another faster vehicle, which is traveling at the speed limit, catches up with the controlled vehicle in the same lane, see Fig. 3.4a. According to the lane-change strategy, the controlled vehicle returns to the original lane. In the simulation example, the reference speed changes according to the speed limit and the vehicle is in the beginning of a downhill section of the journey. Signals in the lane-change maneuver and weighting factors both in the current lane and the adjacent lane are shown in Fig. 3.5.

Figure 3.5a shows the speed of the vehicle. In the first section, the controlled vehicle is traveling at constant 110 km/h velocity when it catches up with a preceding vehicle, which is traveling at 90 km/h. Thus, weight W is increased to avoid the collision with the preceding vehicle, see Fig. 3.5f. The increase of W is induced by the velocity of the preceding vehicle $v_{lead} = 90$ km/h and the predefined safe distance $d_{safe} = 50$ m. At the same time $R_{1,new} = 0$, because there is another vehicle in the adjacent lane, see Fig. 3.5g. When the vehicle in the adjacent lane overtakes the controlled vehicle, $R_{1,new} = R_{1,max}$, in which $R_{1,max}$ is set by the driver, see at station 200 m. Since the speed of this vehicle is greater than that of the controlled vehicle speed, $W_{new} = 0$ is set in the adjacent lane, see Fig. 3.5h. Since the safe distance in the new lane cannot be guaranteed, the overtaking maneuver is not initiated. The overtaking and the lane change begin only at station 310 m. When the overtaking maneuver has been carried out, the current lane will be the new one and the adjacent lane will be the original one. Thus, W in the current lane is reduced to zero and W_{new} in the original lane is increased to 1, because the distance between the controlled vehicle and the overtaken vehicle is decreased, see Fig. 3.5f and h. Furthermore, the velocity is increased in order to achieve the designed velocity, see Fig. 3.5a and the throttle signal Fig. 3.5b.

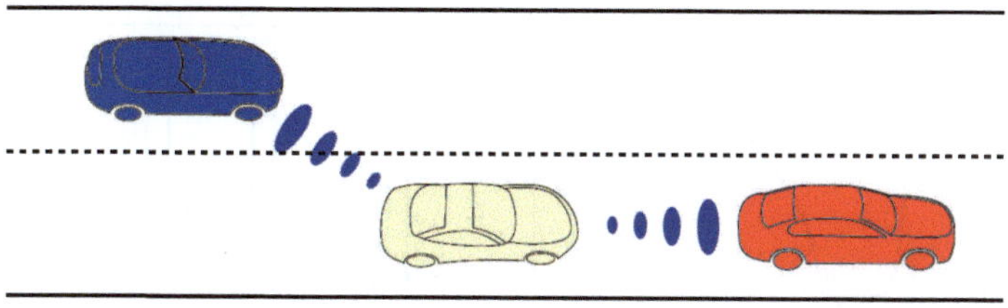

(a) Conflict between the controlled vehicle and the preceding one

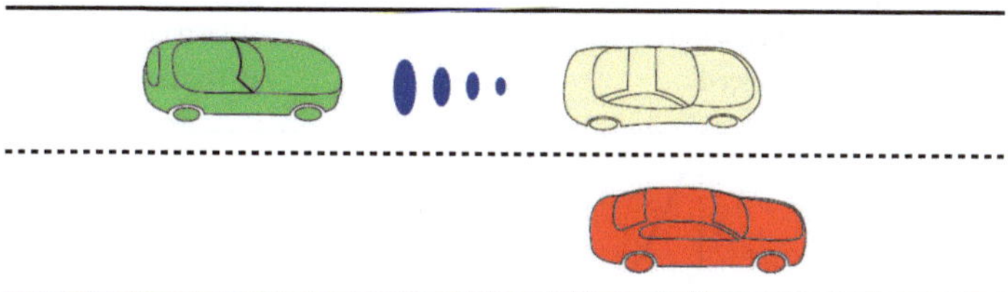

(b) Conflict between the controlled vehicle and the follower one

Fig. 3.4 Illustration of the conflict situations

In the second part of the simulation, the follower vehicle catches up with the controlled vehicle as shown in Fig. 3.5d. When the distance between the controlled vehicle and the follower vehicle decreases to the safe distance, weight R_1 is reduced in order to accelerate the controlled vehicle, see Fig. 3.5e at the station 360 m. Thus, the speed of the controlled vehicle is increased until 125 km/h, as shown in Fig. 3.5a. During the acceleration, the vehicle of the adjacent lane is overtaken, which is shown by the signal W_{new} at station 470 m in Fig. 3.5h. Since the controlled vehicle is faster than the overtaken vehicle, $R_{1,new}$ in the adjacent lane is set to $R_{1,max}$, see Fig. 3.5g. Based on this signal, the lane-change strategy suggests that the controlled vehicle should return to the original lane. When the safety distance from the overtaken vehicle is reached, the overtaking maneuver has been finished and the controlled vehicle is driven back to the lane, see station 600 m. Thus, the distance behind the vehicle increases, while the speed of the vehicle and the throttle are reduced significantly, see Fig. 3.5a and f. Simultaneously, $R_1 = R_{1,max}$ in the new lane and $R_{1,new} = 0$ in the adjacent lane are set.

3.4.3 A Complex Simulation Scenario

In the following, the proposed method is demonstrated through a complex simulation scenario. In the simulation, the high-fidelity vehicle dynamics of the CarSim software is used while the controller implementation is carried out in Simulink. The selected vehicle is a Minivan with the mass of 2037 kg. The speed and lane selection of the controlled vehicle are based on the proposed look-ahead strategy, in which the

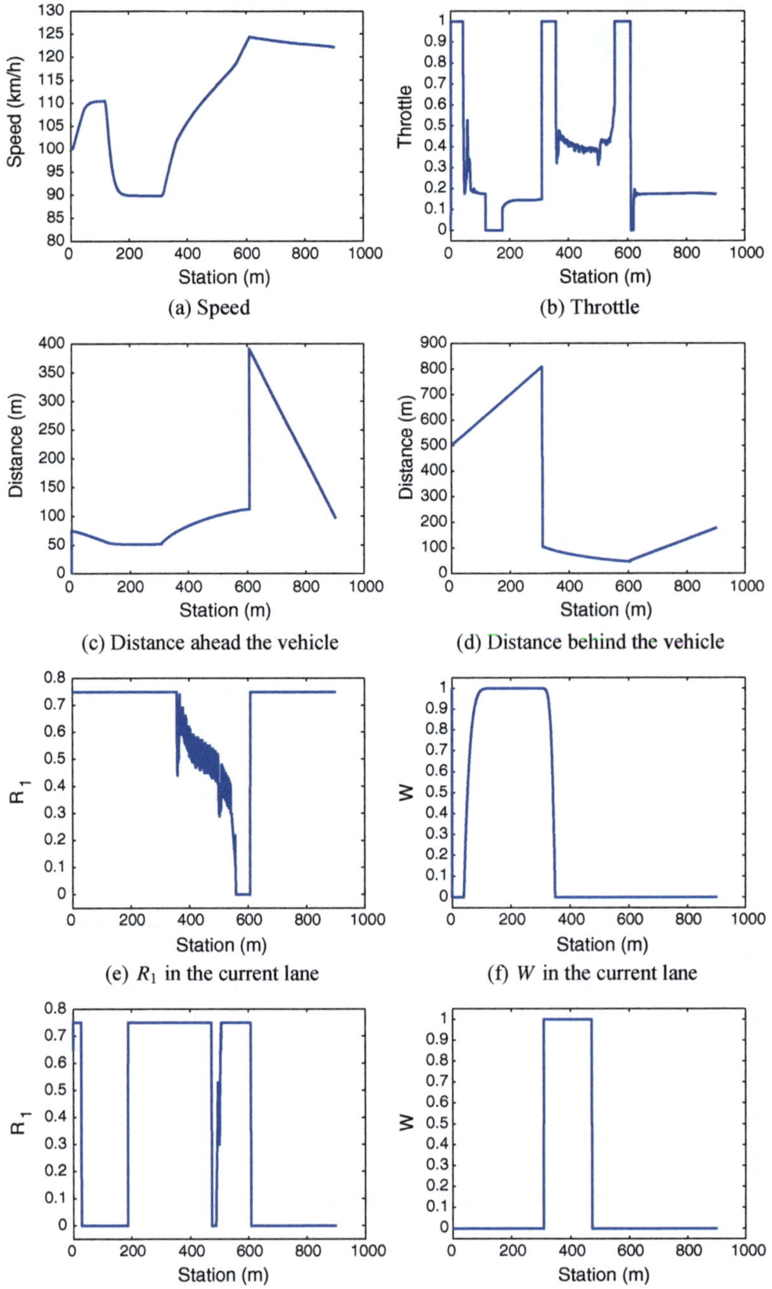

(a) Speed (b) Throttle

(c) Distance ahead the vehicle (d) Distance behind the vehicle

(e) R_1 in the current lane (f) W in the current lane

(g) $R_{1,new}$ in the adjacent lane (h) W_{new} in the adjacent lane

Fig. 3.5 Signals in the lane-change maneuver

information about the surrounding vehicles is considered. In the simulation, the look-ahead horizon of the vehicle is 1000 m with 10 section points.

The simulation contains three different overtaking maneuvers. First, the controlled vehicle overtakes the preceding vehicle without further disturbing vehicles, see Fig. 3.6. Second, the controlled vehicle catches up the slower preceding vehicle, see Fig. 3.7. However, there is a fast disturbing vehicle in the adjacent lane. Since the distance between the controlled vehicle and the disturbing vehicle is smaller than $d_{safe} = 40$ m, the overtaking cannot be performed. Thus, the controlled vehicle must adapt to the preceding slow vehicle, which results in the reduction of the speed. When the distance between the disturbing vehicle and the controlled vehicle

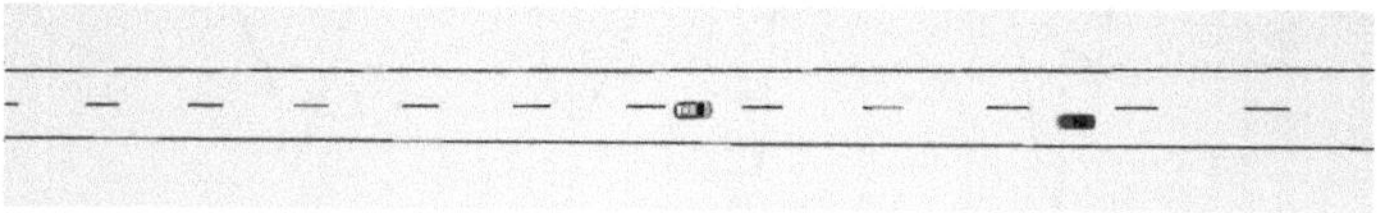

Fig. 3.6 First overtaking maneuver (controlled vehicle: silver, preceding vehicle: blue)

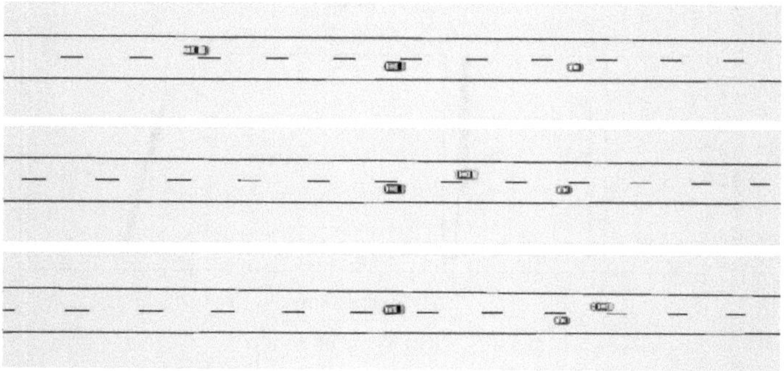

Fig. 3.7 Second overtaking maneuver (controlled vehicle: silver, preceding vehicle: white, disturbing vehicle: yellow)

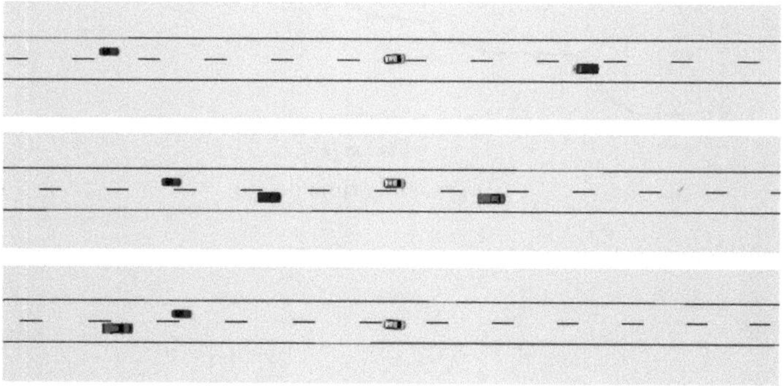

Fig. 3.8 Third overtaking maneuver (controlled vehicle: silver, preceding vehicles: red and green, disturbing vehicle: blue)

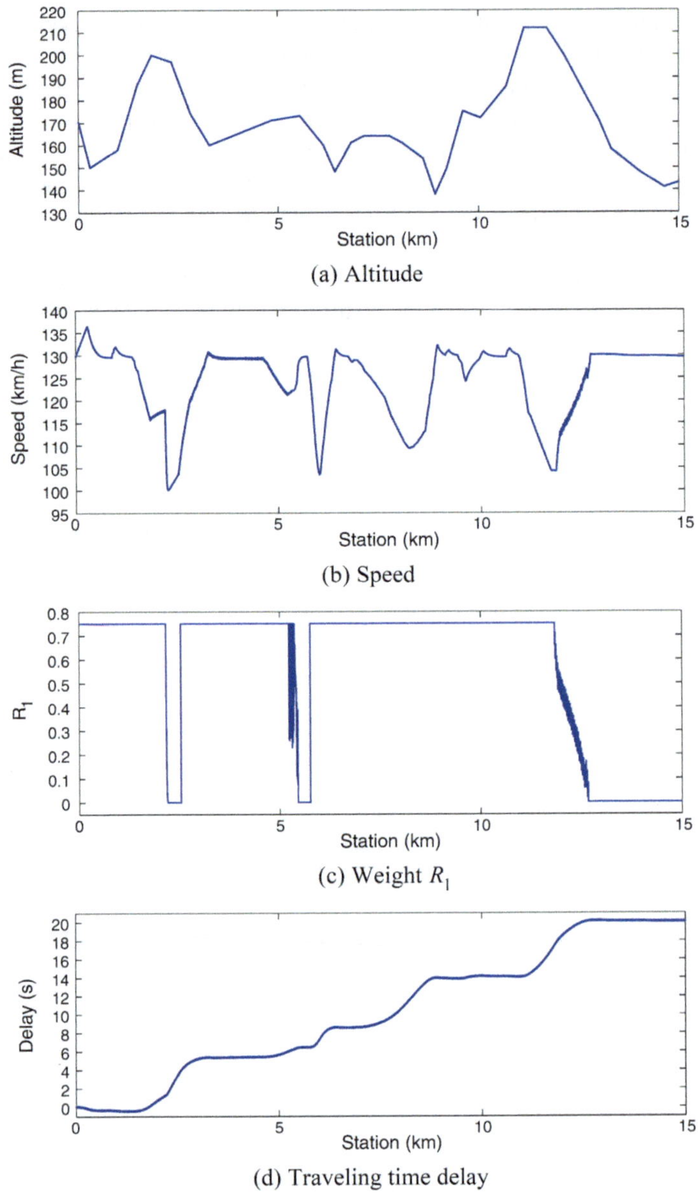

(a) Altitude

(b) Speed

(c) Weight R_1

(d) Traveling time delay

Fig. 3.9 Simulation results

is higher than d_{safe}, the overtaking can be performed, as shown in Fig. 3.7. In the third maneuver, the controlled vehicle overtakes slow vehicles, moving to the adjacent lane, see Fig. 3.8. However, a fast disturbing vehicle catches up the controlled vehicle. Due to the slow vehicles, the controlled vehicle cannot move back to the original lane. Therefore, the controlled vehicle must adapt to the follower vehicle with the increasing of the speed. When the distance between the controlled vehicle and the leader slow vehicle is higher than d_{safe}, the controlled vehicle moves back to its original lane, and the energy-efficient cruising can be selected.

Figure 3.9a presents the altitude of the road. In the simulation example, the real data of a 15-km-long segment of the Hungarian M1 highway is used. The altitude information is resulted from the database of Google Earth. The speed limit is 130 km/h on the entire segment. Figure 3.9a shows that the highway contains several uphill and downhill sections. Therefore, the look-ahead control can be efficient from the aspect of energy-optimal cruising. Figure 3.9b illustrates the speed variation of the vehicle. The impact of the terrain characteristics on the speed profile shows clearly the section, e.g., between 6.5 km and 9 km. Furthermore, the speed profile is influenced by the further vehicles of the traffic and the time management strategy.

Figure 3.9c demonstrates the selection of R_1. The reduction of R_1 between 2.3 km and 2.7 km is resulted by the second overtaking maneuver. It also can be seen that the selection of R_1 modifies the speed of the vehicle (Fig. 3.9b), when the controlled vehicle is adapted to the speed preceding slower vehicle v_{lead}. Moreover, between the segments 5.2–6 km the third overtaking maneuver is performed. In this case, the weight R_1 must be reduced to adapt to the speed of the follower vehicle, which is reached at 5.6 km. Finally, the R_1 is modified from the segment point 12 km to guarantee the predefined schedule delay. In the simulation, the maximum delay is 5%, which is 20 s. The increase of the delay is shown in Fig. 3.9d. It presents that the maximum delay t_{max} is guaranteed through the variation R_1.

The simulation results show that the lane-change strategy of the look-ahead control operates effectively. The necessity of the lane changes is decided and the safe realization is carried out. The simulation example has shown that by tuning the predicted speed conflicts and emergencies with other vehicles can be reduced.

References

Ambuhl D, Guzzella L (2009) Predictive reference signal generator for hybrid electric vehicles. IEEE Trans Veh Technol 58(9):4730–4740

Asadi B, Vahidi A (2011) Predictive cruise control: Utilizing upcoming traffic signal information for improving fuel economy and reducing trip time. IEEE Trans Control Syst Technol 19(3):707–714

Barth A, Franke U (2008) Where will the oncoming vehicle be the next second? In Proceedings of the IEEE intelligent vehicles symposium, pp 1068–1073

Broadhurst AE, Baker S, Kanade T (2004) A prediction and planning framework for road safety analysis, obstacle avoidance and driver information. In: 11th World congress on intelligent transportation systems, vol 1, pp 1–12

Gáspár P, Németh B (2015) Design of look-ahead cruise control using road and traffic conditions. In: American control conference, Chicago, USA, pp 3447–3452

Németh B, Gáspár P (2013b) Design of vehicle cruise control using road inclinations. Int J Veh Auton Syst 11(4):313–333

Németh B, Gáspár P (2014) Optimised speed profile design of a vehicle platoon considering road inclinations. IET Intell Transp Syst 8(3):200–208

Saerens B, Rakha H, Diehl M, den Bulck EV (2013) A methodology for assessing eco-cruise control for passenger vehicles. Transp Res Part D 19:20–27

van Keulen T, de Jager B, Foster D, Steinbuch M (2010) Velocity trajectory optimization in hybrid electric trucks. In: American Control Conference, pp 5074–5079

Chapter 4
Design of Predictive Cruise Control for Safety Critical Vehicle Interactions

Introduction and Motivation

The interactions of autonomous vehicles in smart cities are an important research field in autonomous vehicles. Since in urban areas a large number of intersections are found, there is a huge potential in the adequate control of vehicles approaching the intersections. The coordinated control of autonomous vehicles provides a more flexible solution for vehicle interactions in intersections and is able to improve effectiveness and, simultaneously, the safety of the traffic system.

Figure 4.1 illustrates autonomous vehicles crossing intersections. A conventional intersection with traffic signs is illustrated in Fig. 4.1a. In the scenario, three vehicles are approaching the intersection. *Vehicle 1* has right of way against the other two vehicles since it is traveling a higher order road and the other two vehicles have give way signs. *Vehicle 2* intents to go straight ahead, while *Vehicle 3* intents to turn left. In the scenario, both *Vehicle 2* and *Vehicle 3* must be decelerated and stopped if necessary. It can lead to increasing energy and fuel consumptions and loss of time.

Without traffic signs (Fig. 4.1b) in the scenario, it is possible to modify the motion profile of the vehicles to achieve minimum traveling time and energy consumption. However, it is not a trivial question which ordering of the vehicles in the intersection is the best choice, as it is illustrated by the following example. It is assumed that *Vehicle 1* has an internal combustion engine, and its motion is slow, while *Vehicle 2* is a fast electric vehicle. Moreover, *Vehicle 3* is a truck with high mass. The ordering of the vehicles may vary, depending on the objective function.

- If the goal is the minimization of traveling time, then the ordering *Vehicle 2-Vehicle 1-Vehicle 3* is the best choice, which means that the current speeds of the vehicles have high relevance.
- If energy loss must be minimized, the ordering *Vehicle 3-Vehicle 1-Vehicle 2* is optimal, because the heavy vehicle is not stopped unnecessarily.

P. Gáspár and B. Németh, *Predictive Cruise Control for Road Vehicles Using Road and Traffic Information*, Advances in Industrial Control,
https://doi.org/10.1007/978-3-030-04116-8_4

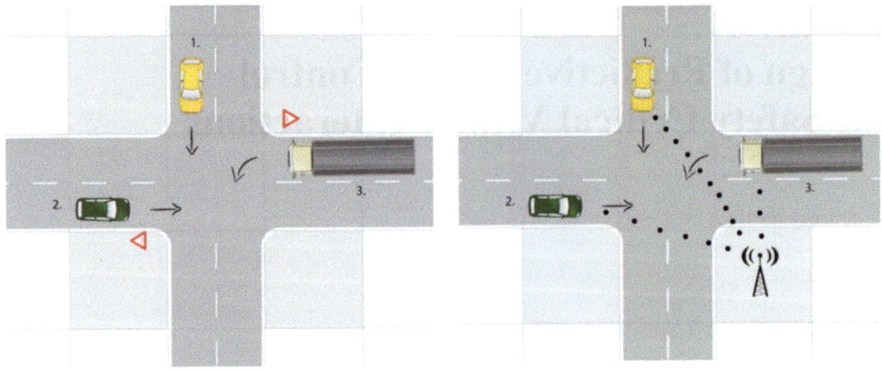

(a) Intersection with traffic signs (b) Intersection w/o traffic signs

Fig. 4.1 Illustration of intersection scenarios

- However, if the reduction of emission in the intersection is the objective, the ordering *Vehicle 1-Vehicle 3-Vehicle 2* is the appropriate selection to avoid the start/stop of the conventional vehicle.

Since there is a contradiction between the various performances, in the control of the autonomous vehicles, a balance between them must be guaranteed.

The adequate control of the motions of autonomous vehicles in intersections has several advantages. In a real traffic scenario, autonomous vehicles are traveling together with human-driven vehicles and pedestrians. It is a necessary condition to set clearly the traffic rules, which determine the motions of the various types of travelers. Consequently, these traffic rules reduce the possibilities of autonomous vehicles and, moreover, the behavior of humans must be incorporated in the control of the intersection. The detection and prediction of human behavior are extremely difficult and complex problems, see Li et al. (2016), Aoude et al. (2012).

Moreover, the architecture in the control strategy of the autonomous vehicles in intersections should be varied. A centralized coordination of the vehicles in the intersection results in an efficient operation of the traffic system. However, it requires enhanced infrastructure, e.g., Vehicle-to-vehicle and vehicle-to-infrastructure communications, see Wuthishuwong and Traechtler (2013). Moreover, the autonomous vehicles must have a decision logic in them with which the interactions are handled. Thus, the control problem of smart intersections has both centralized and individual components.

Several papers have produced various results concerning intersections. Some of them are listed in the following. A framework for the intersection control, which is based on the queuing theory, was presented by Tachet and Santi (2016). Through this framework, a capacity-optimal slot-based intersection management system for two road crossing configurations was presented. A model predictive control based intersection control using a centralized approach for two vehicles was presented by Riegger et al. (2016). The quadratic programming method is a possibility for real-time implementation compared to the convex optimization using space coordinates, see Murgovski et al. (2015). Collision avoidance is guaranteed through the definition of regions in the intersection with special rules which reduce the complexity of the problem, but the conservativeness of the solution increases. An autonomous intersection management system in which the connection of intersections is handled by a multi-agent viewpoint was presented by Dresner and Stone (2008), Hausknecht et al. (2011). This solution provides a possibility to quickly reverse individual lanes in response to rapidly changing traffic conditions. Another multi-agent solution, which is based on a heuristic optimization algorithm, was presented by Zohdy and Rakha (2012). The objective of the research is to reduce total time delay for the entire intersection while collisions are prevented. A mixed-integer linear programming based method for the coordination of vehicles in the intersection was published by Fayazi et al. (2017). The goal of the control is to find an arrival schedule of the vehicles, which ensures safety while it reduces the number of stops and intersection delays.

Furthermore, the novel infrastructure elements have a role in the coordination of autonomous vehicles in smart intersections. Through Vehicle-to-Infrastructure (V2I) communication technologies, the smart elements are able to transmit information to the intelligent vehicles, see e.g., Gardner et al. (2009). As an application of the traffic information, in Asadi and Vahidi (2011) a predictive cruise control was proposed, which is able to consider upcoming traffic signal information to improve fuel economy and reduce traveling time. The algorithm uses radar and traffic signals to optimize traveling speed. However, in a fault scenario of an infrastructure element, these criteria cannot be guaranteed by the traffic control system. Therefore, the intelligent vehicles must handle the critical situations, while the energy-efficient motion is guaranteed by the vehicle cruise control systems. It requires the estimation of the other participants in the traffic scenario and an individual decision strategy about the ego vehicle motion. Estimation methods of surrounding vehicle motion prediction for safety critical traffic applications were proposed in Barth and Franke (2008), Broadhurst et al. (2004).

Handling traffic lights and infrastructure element faults is a special task in intelligent vehicle control design. The concept of fault-tolerant and energy-efficient speed profile design is proposed in the following. Figure 4.2 illustrates that the vehicle receives information from the traffic light infrastructure element about the unexpired time of the current signal (red/amber/green) and the operation of the light. If the operation of the traffic light is fault-free, then the speed profile based on the unexpired time is designed, see Németh et al. (2013). However, if there is a fault in the operation, other information sources are required to guarantee safe vehicle motion, such as the motion of the other vehicles (distance, speed, acceleration). Information can be supplied by vehicle-to-roadside and vehicle-to-vehicle communication, as proposed in Kesting et al. (2007) and Festag et al. (2008). Moreover, the road inclination and speed limit data are also considered to be known. Since the route of the vehicle is considered as known, the loading of this information requires the measurement of the current position, e.g., using GPS. In the case of a traffic light fault, from safety aspects, two further scenarios can be distinguished. First, the con-

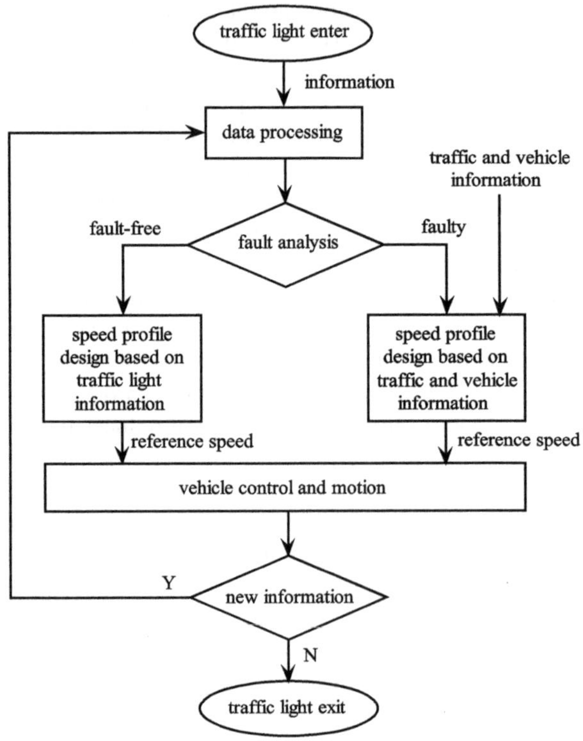

Fig. 4.2 Flowchart of the speed profile design strategy

trolled vehicle moves on a primary route and none of the vehicles has priority in the intersection. In this scenario, the motion of the vehicle is not influenced by the other vehicles. Safe driving must be guaranteed by the vehicles on the non-primary routes by observing the traffic rules. The speed profile is designed according to Németh et al. (2013). Second, the controlled vehicle moves on a non-primary route. It means that the motion of the vehicles on the primary routes must be considered in the design of the speed profile. The vehicles which have influence on the controlled vehicle speed profile are called disturbance vehicles. In this scenario, the speed profile must be designed to avoid the critical safety conflicts between the controlled and disturbance vehicles.

4.1 Strategy of Vehicle Control in Intersections

It has been shown that the control of intelligent vehicles in intersections depends on several factors. However, the proposed predictive control algorithm of Sect. 2.2 provides an appropriate formulation, in which the energy and the traveling time requirements can be handled together. In further examination of the intersection problem, it is required to find the speed profile of the vehicles, which is able to guarantee a balance between energy consumption and traveling time through the performance weight R_1. Moreover, the designed speed trajectories must guarantee the safe motion of vehicles, which means the avoidance of collisions. It is guaranteed by the selection of W. It means that the speed of the vehicles with predictive cruise control can be influenced through the weights R_1 and W.

The fundamental idea of vehicle control in intersections is to find the maximum set of vehicles in which the vehicles are able to be in cooperation. Cooperation means that the safe motion of the controlled vehicles in the set can be guaranteed without a stopping command. In the set, all of the controlled vehicles can be moved through R_1 and $W = 0$. The controlled vehicles, which are out of the set, can be influenced through R_1 and $W \neq 0$ to guarantee their safe motion. The role of optimization in the intersection is to find the maximum set of vehicles in which cooperation can be guaranteed.

In real traffic scenarios of intersections, the human-driven vehicles and the intelligent vehicles with predictive cruise control travel together. Due to the human drivers, it is necessary to impose some regulations by which the motion of their vehicles can be controlled. Therefore, the intelligent vehicles must adapt to the regulations and these constraints must be incorporated in the optimization problem. During the design of intelligent vehicle control strategy, the estimation of the motion of the human-driven vehicles is carried out. It results in the predicted speed profile of the conventional vehicles, which is based on the available signals, e.g., current position, speed, and acceleration, see Sect. 4.2.

During the optimization process, the human-driven vehicles can be a part of the maximum set of vehicles in which the vehicles are able to be in cooperation. It means that the speed profile can be found for intelligent vehicles by which collision with human-driven vehicles can be avoided and stopping commands are not used for the vehicles in the set. Regarding conventional vehicles, it is assumed that humans keep the traffic regulations. Similarly, a confidence must be created in the human participants that the intelligent vehicles will also keep regulations and collisions are avoided in the intersection. Therefore, an increased safe distance between the conventional and the intelligent vehicles is guaranteed.

The computation of the maximum set of cooperating vehicles requires the definition of a quantity by which the size of the set is defined. It influences the order of the vehicles in the intersection and the number of the cooperating vehicles. Moreover, this quantity may represent the importance of vehicles regarding the transportation system or some social aspects. For example, quantity can be defined as the number of passengers in the vehicles, which means that public transport buses may have an increased weight. Another example is the presence of high-priority vehicles in the traffic (e.g., ambulances, fire engines), whose passage must absolutely be guaranteed, see e.g., Dang et al. (2015), Smith et al. (2010), Nellore and Hancke (2016). In this strategy, a quasi-kinetic energy is defined as the quantity of the set. The energy of the vehicle i is formed as follows:

$$E_i(t) = (1 - \eta_i)\frac{1}{2}m_i v_i^2(t), \tag{4.1}$$

where m_i is the virtual mass of the vehicle and $v_i(t)$ is its current speed. The parameter η_i is in relation with the efficiency of the driveline recuperation. The parameters of Eq. (4.1) are chosen as follows.

- The virtual mass m_i is in relation with the mass of the vehicle. Generally, m_i is equal to the nominal mass of the vehicle, e.g., in passenger cars. However, in the case of public transport buses, m_i may also include the mass of the passengers. In this way, the number of passengers in the vehicle can be considered in the optimization problem. Since m_i represents an important factor, it is sufficient to estimate the additional mass, e.g., the number of passengers is multiplied by an average unit mass. Furthermore, the importance of the high-priority vehicles can be scaled through a huge virtual mass.
- The current speeds $v_i(t)$ of the vehicles are assumed to be available. Their role is to consider the fact that the stopping of a high-speed vehicle results in larger energy loss, compared to a vehicle with slow motion.
- Since there may be several hybrid and electric vehicles in the traffic system, their energy loss can be reduced through recuperation. Parameter η_i is the average efficiency of recuperation in the driveline. It shows that the energy loss of a hybrid or electric vehicle is smaller compared to a conventional driveline with an internal combustion engine.

During the optimization problem of the intersection, the sum of the quasi-kinetic energy of the set must be maximized:

$$\max_{\omega_{k,l} \in \Omega} \sum_{i=1}^{k} E_i(t), \tag{4.2}$$

where Ω is the set of all vehicles in the region of interest around the intersection and in $\omega_{k,l}$ subset k is the number of the vehicles within l subset. Since several $\omega_{k,l}$ subsets of the vehicles from Ω can be generated, it is necessary to order them according to $\sum_i E_i(t)$. Then, it is necessary to examine whether the vehicles of $\omega_{k,l}$ are able to be in cooperation. The result of the optimization is the maximum $\omega_{k,l}$ in which the cooperation is available.

For example, a location in which there are four intelligent vehicles with predictive cruise control is considered an intersection. The set combinations are the following:

- all of the vehicles are in cooperation $\omega_{4,A}$;
- three vehicles are in cooperation, which means $\binom{4}{3} = 4$ number of combinations, such as $\omega_{3,A}$, $\omega_{3,B}$, $\omega_{3,C}$, $\omega_{3,D}$;
- two vehicles are in cooperation, which results in $\binom{4}{2} = 6$, such as $\omega_{2,A}$, $\omega_{2,B}$, $\omega_{2,C}$, $\omega_{2,D}$, $\omega_{2,E}$, $\omega_{2,F}$;
- it is not possible to find cooperating vehicles, which leads to sets with one element. In this case there are $\binom{4}{1} = 4$ combinations $\omega_{1,A}$, $\omega_{1,B}$, $\omega_{1,C}$, $\omega_{1,D}$.

Thus, the number of the candidate maximum set is altogether $\binom{4}{4} + \binom{4}{3} + \binom{4}{2} + \binom{4}{1} = 15$. Then, it is necessary to order the sets $\omega_{k,l}$ according to their energy $\sum\limits_i E_i(t)$.

In summary, the process of the speed profile design of the vehicles in the intersection is the following:

1. the determination of the region of interest around the intersection,
2. the prediction of the speed profile of the human-driven vehicles,
3. the determination of the quasi-kinetic energies of the vehicles,
4. the ordering of the candidate subset of the cooperating vehicles,
5. the selection of the maximum set of the cooperating vehicles through prediction and optimization algorithms, and
6. the ordering and speed profile design of the vehicles which are out of the set.

4.2 Motion Prediction of Vehicles in the Intersection

The design of speed profiles for intelligent vehicles requires the prediction of their own motion and the motion of the human-driven vehicles. In the following, the prediction processes of the vehicles are introduced.

4.2.1 Motion Prediction of Human-Driven Vehicles

In the prediction process of human-driven vehicles, several assumptions are made. First, it is assumed that a controlled vehicle has information about the speed and acceleration of the disturbance vehicles ($\dot{\eta}$, $\ddot{\eta}$) and about its current distance from the intersection e_η. Second, the disturbance vehicles accelerate evenly until they reach the speed limit. When a human-driven vehicle reaches the speed limit $v_{ref,j}$, it does not accelerate further; thus in the oncoming sections, the predicted speeds of the vehicle are $v_{ref,j}, \ldots, v_{ref,n}$.

The considered speed profile of the conventionally driven vehicle is illustrated in Fig. 4.3. The following distance and acceleration equations describe the speed profile:

$$e_\eta = \dot{\eta} t_{acc} + \frac{v_{ref,int} - \dot{\eta}}{2} + v_{ref,int}(t_{int} - t_{acc}) \tag{4.3}$$

$$\ddot{\eta} = \frac{v_{ref,int} - \dot{\eta}}{t_{acc}}, \tag{4.4}$$

where $v_{ref,inf}$ is the speed limitation on the route of the vehicle. t_{acc} is the duration of the acceleration, while t_{int} is the time required to reach the intersection. t_{int} is expressed from Eq. (4.4) as

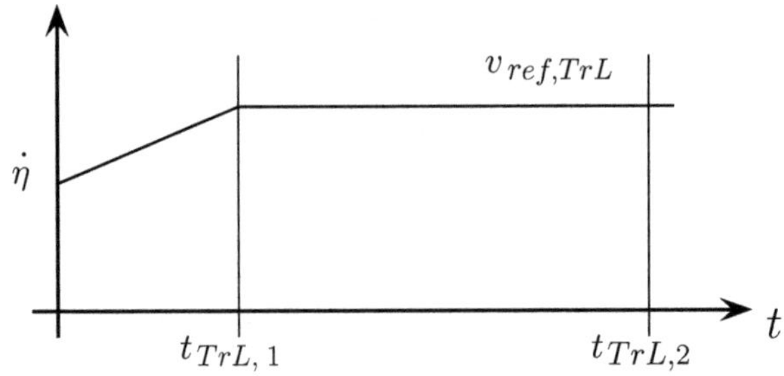

Fig. 4.3 Speed profile of the human-driven vehicle

$$t_{int} = \frac{v_{ref,int} - \dot{\eta}}{\ddot{\eta}} + \\ + \frac{1}{v_{ref,int}} \left[e_\eta - \dot{\eta} \frac{v_{ref,int} - \dot{\eta}}{\ddot{\eta}} - \frac{(v_{ref,int} - \dot{\eta})^2}{2\ddot{\eta}} \right]. \quad (4.5)$$

Since time t_{int} represents the critical situation when the human-driven vehicle reaches the intersection, this time value is considered at the computation of the controlled vehicle motion.

4.2.2 Speed Prediction of the Controlled Vehicle

Moreover, it is also required to provide a speed prediction method on the controlled vehicles. The calculation is the same as it was in the case of the motion of the follower vehicle, which is analyzed in Sect. 3.2.

The predicted speed of the vehicle at section point n is calculated in the following steps. Using Eq. (3.6), ϑ can be rewritten as

$$\vartheta = v_{ref,0}^2 (1 - R_1) + R_1 \bar{\vartheta}, \quad (4.6)$$

where $\bar{\vartheta}$ contains the value of ϑ calculated with energy-efficient prediction weights $\bar{Q}, \bar{\gamma}_i$. From (3.7), the square of the reference speed λ is calculated based on the predicted road information.

$$\lambda^2 = v_{ref,0}^2 (1 - R_1) + R_1 \bar{\lambda}^2. \quad (4.7)$$

Using (3.10), the square of the predicted estimated speed of the vehicle at section point n is as follows:

$$\dot{\xi}_n^2 = (R_1 \mathcal{N}_1 + \mathcal{N}_2) R_1 + (1 - R_1) v_{ref,n}^2, \quad (4.8)$$

which depends on the weight R_1. In the formula, $\mathcal{N}_1$ is independent of the number of section points ahead, while $\mathcal{N}_2$ contains the road grade information of each section.

The aim of speed prediction is to determine the position of the controlled vehicle at various times. The controlled vehicle moves from point ξ_0 to ξ_1 covering the distance s_1 while the traveling time in this section is Δt_1. In the optimization method, it is assumed that although the acceleration of the vehicle may change in the different intervals, within an interval acceleration is constant. Thus, the traveling time in the first interval is expressed as $\Delta t_1 = 2s_1/(\dot{\xi}_1 + \lambda)$, where λ and $\dot{\xi}_1$ are from Eqs. (3.7) and (3.10), respectively. The traveling time between points ξ_1 and ξ_2 is expressed similarly to $\Delta t_2 = 2s_2/(\dot{\xi}_2 + \dot{\xi}_1)$. Using the computed time values, a vector is formed as $\Delta T = \left[\Delta t_1 \dots \Delta t_n\right]^T$. Moreover, the positions of the section points are also arranged in a vector form such as $\Xi = \left[\xi_1 \dots \xi_n\right]^T$.

4.3 Optimal Speed Profile Design

The optimization process is based on the modification of R_1 weight of all vehicles which is in a candidate set of cooperating vehicles. The role of R_1 is to guarantee a balance between the energy-optimal and the minimum traveling time motion profiles. Similarly to the cruising strategy in the overtaking scenarios (see Chap. 3), it is necessary to find the maximum R_1 of each vehicle, by which the safe motion of other vehicles in the intersection can be guaranteed.

However, if a problem occurs in the intersection, it is necessary to design R_1 weight for several vehicles simultaneously. Therefore, in the following, $R_{1,i}(t)$ represents the performance weight of vehicle i at time t. Moreover, it is necessary to define an objective function by which the importance of the controlled vehicles can be scaled. The previously defined $E_i(t)$ may be a good choice, because this value has the same role. Thus, the objective function of the optimization is

$$\max_{R_{1,i}(t) \in [0; R_{1,max}]} \sum_{i \in \omega} E_i(t) R_{1,i}(t), \tag{4.9}$$

where $R_{1,max}$ is the upper limit of $R_{1,i}(t)$.

The objective of the optimization is constrained by some safety criteria which guarantee the avoidance of collisions between the vehicles in set ω. Regarding the constraints, a safe distance s_{safe} between the vehicles is predefined. The value s_{safe} depends on the speed limits $v_{ref,i}$ in the intersection and the number of roads connected to the intersection. The constraint between vehicle i and j represents the following. If vehicle j reaches the intersection at the predicted time t_j, vehicle i must be out of the intersection area at least by the distance s_{safe}. The form of the constraints is defined as

$$e_i - \xi_i(t_j) > s_{safe} \qquad \forall i \neq j, \tag{4.10}$$

where i represents vehicle i and e_i is the current distance of vehicle i from the intersection. $\xi_i(t_j)$ is the predicted distance of vehicle i from its current position at time t_j.

Thus, the optimization problem is formed as

$$\max_{R_{1,i}(t)\in[0;R_{1,max}]} \sum_{i\in\omega} E_i(t)R_{1,i}(t), \tag{4.11}$$

such that

$$e_i - \xi_i(t_j) > s_{safe}, \qquad \forall i \neq j.$$

The solution of the optimization problem Eq. (4.11) requires some prediction processes. It is necessary to select a candidate $R_{1,j}$ value for each vehicle j and to provide a prediction as follows. Using $R_{1,j}$, the vectors of the prediction for vehicle j $\varXi$, $\varDelta T$ are computed, see Sect. 4.2. Using e_j current distance of vehicle j from the intersection, t_j is computed as

$$t_j = interp(\varXi_j, \varDelta T_j, e_j). \tag{4.12}$$

Then, $\xi_i(t_j)$ can be determined for all vehicles $i \neq j$.

- In the case of intelligent-controlled vehicles, $\varXi_i$ and $\varDelta T_i$ are computed (Sect. 4.2). Then, $\xi_i(t_j)$ is computed as

$$\xi_i(t_j) = interp(\varDelta T_i, \varXi_i, t_j). \tag{4.13}$$

- For human-driven vehicles, $\xi_i(t_j)$ is computed through the prediction of the motion profile, see Fig. 4.3.

Thus, the constraints regarding vehicle j and i are computed by using Eq. (4.10).

As an example, in the case of three intelligent vehicles, the following constraints are determined:

- $j = 1$: For a selected value of $R_{1,1}$, the vectors $\varXi_1$, $\varDelta T_i$ are computed, which results in t_1 using the measurement e_1, see Eq. (4.12). Moreover, (3.10) results in $\xi_2(t_1)$, $\xi_3(t_1)$ through the computation of $\varDelta T_2$, $\varDelta T_3$, $\varXi_2$, $\varXi_3$. The constraints are $e_2 - \xi_2(t_1) > s_{safe}$ and $e_3 - \xi_3(t_1) > s_{safe}$.
- $j = 2$: Similarly, in the case of a given $R_{1,2}$ the value of t_2 is computed. It yields the constraints $e_1 - \xi_1(t_2) > s_{safe}$ and $e_3 - \xi_3(t_2) > s_{safe}$, as presented above.
- $j = 3$: Finally, $R_{1,3}$ leads to the computation of t_3, which yields the constraints $e_1 - \xi_1(t_3) > s_{safe}$ and $e_2 - \xi_2(t_3) > s_{safe}$.

Thus, the optimization problem in this example is formed as

$$\max_{R_{1,i}(t)\in[0;R_{1,max}]} \sum_{i=1}^{3} E_i(t)R_{1,i}(t) \tag{4.14}$$

such that

$$e_2 - \xi_2(t_1) > s_{safe}, \quad e_3 - \xi_3(t_1) > s_{safe}, \quad e_1 - \xi_1(t_2) > s_{safe},$$
$$e_3 - \xi_3(t_2) > s_{safe}, \quad e_1 - \xi_1(t_3) > s_{safe}, \quad e_2 - \xi_2(t_3) > s_{safe}.$$

In practice, the solution of the optimization processes Eq. (4.11) may require a great deal of computation effort. Therefore, in the computation the result of the previous optimization step is used, such as $R_{1,1}^{old}$, $R_{1,2}^{old}$, $R_{1,3}^{old}$. The new solution $R_{1,i}^{new}$ is searched for in the interval $[\max(R_{1,1}^{old} - \alpha, 0), \min(R_{1,i}^{old} + \alpha, R_{1,max})]$, where the interval is divided into n points and α defines the search range. The purpose of this procedure is to guarantee that the complexity of the optimization method is reduced and, thus, the method can be applied in practice.

4.4 Simulation Results

The predictive speed profile design of the vehicles in the intersection is illustrated through two simulation examples. In the first scenario, three autonomous vehicles are in the region of interest around the intersection. Thus, it is possible to control the motion of three vehicles simultaneously. In the second example, two vehicles are found in the intersection: a human-driven vehicle and an autonomous vehicle. This scenario illustrates that the cruise control strategy is able to handle the mixed traffic with various vehicles.

4.4.1 Interaction of Autonomous Vehicles

In the scenario, there are three intelligent vehicles with predictive cruise control, whose positions and turning intentions are illustrated in Fig. 4.4. Moreover, the speed limit is 50 km/h on the network, but it varies on two roads. Although the intersection is considered to be flat, due to the various speed limits the speed profile of the vehicles must be optimized. In the example, the masses of the vehicles and their respective distances from the center of the intersection ξ_0 are different. The initial speeds of the vehicles are considered to be equal to the speed limit, such as 50 km/h for *Vehicle 1* and *Vehicle 2*, while it is 60 km/h for *Vehicle 3*.

The determination of the maximum set of cooperating vehicles is based on the energy values of the vehicles, as proposed in Sect. 4.1. Considering $\eta_i = 0$ for all vehicles, the energy values in $t = 0$ are $E_1(0) = 244$ kJ, $E_2(0) = 109$ kJ, $E_3(0) = 212$ kJ. Therefore, the ordering of the sets is as follows:

- all vehicles are in cooperation: ω_3 with 565 kJ;

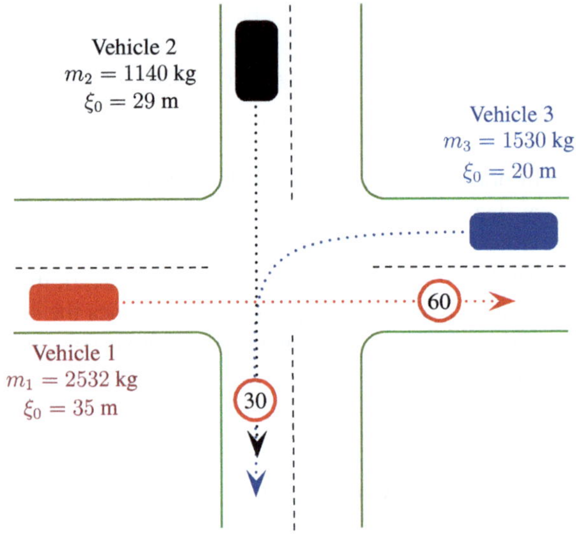

Fig. 4.4 Scenario of the vehicles with their initial positions

- two vehicles are in cooperation: $\omega_{2,1}$ (*Vehicle 1-Vehicle 3*) with 456 kJ, $\omega_{2,2}$ (*Vehicle 1-Vehicle 2*) with 353 kJ, $\omega_{2,3}$ (*Vehicle 2 - Vehicle 3*) with 321 kJ;
- cooperation cannot be reached: $\omega_{1,1}$ (*Vehicle 1*) with 244 kJ, $\omega_{1,2}$ (*Vehicle 3*) with 212 kJ, $\omega_{1,3}$ (*Vehicle 2*) with 109 kJ.

The simulation example shows that *Vehicle 3* is the first in the intersection, *Vehicle 1* is the second, while *Vehicle 2* is the last vehicle, see Fig. 4.5. During the simulation, the safe distance is selected as $s_{safe} = 3$ m, which must be guaranteed in all interactions.

The most important results of the simulations regarding the predictive cruise control strategy are found in Fig. 4.6. The speed values of the vehicles are illustrated in Fig. 4.6a. It shows that the speed profile of *Vehicle 3* is steadily decreasing, while *Vehicle 2* must be stopped to avoid the collision with *Vehicle 1*. The $R_{1,i}$ values are illustrated in Fig. 4.6b. It shows that $R_{1,3}$ has the maximum value during the simulation, which means that *Vehicle 3* is able to cruise with its minimum energy consumption. However, $R_{1,1}$ is reduced between 1.2 s. . .1.7 s, which results in its increased acceleration to guarantee the s_{safe} between *Vehicle 1-Vehicle 2*. *Vehicle 2* has the most changeable $R_{1,2}$ signal, as shown in Fig. 4.6b. Moreover, Fig. 4.6c shows that W for *Vehicle 2* must be modified several times. As a result, the most of the simulation ω_3 can be reached, but in some time periods (e.g., 0.6 s. . .0.9 s, 1.1 s. . .1.3 s, 1.7 s. . .2.6 s) $\omega_{2,1}$ is the maximum set with the coordination of *Vehicle 1* and *Vehicle 3*. However, the proposed strategy is able to guarantee the predefined safe distance between the vehicles.

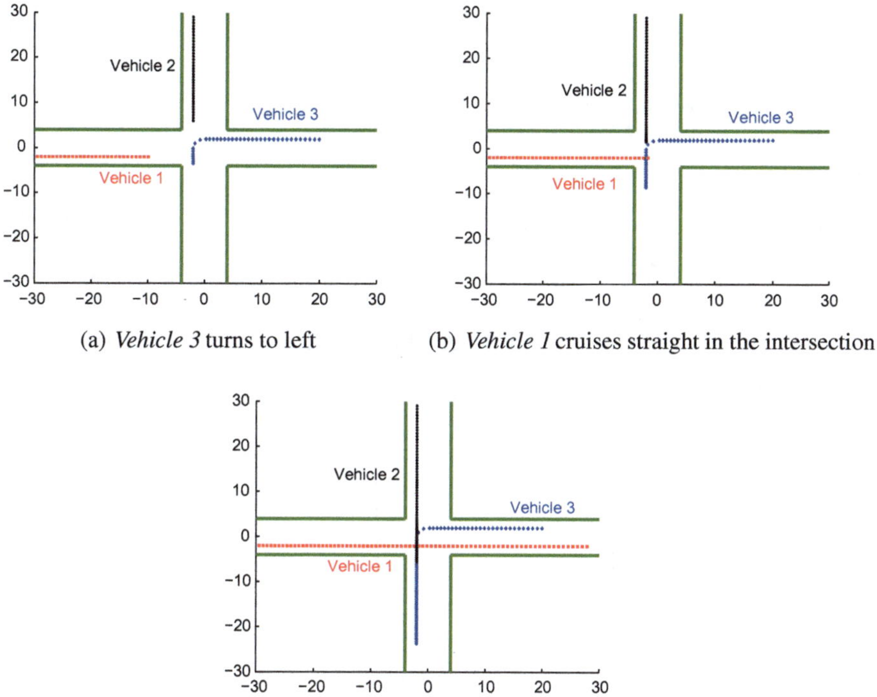

(a) *Vehicle 3* turns to left

(b) *Vehicle 1* cruises straight in the intersection

(c) *Vehicle 2* cruises as the last vehicle

Fig. 4.5 Cruising of the intelligent vehicles in the intersection

4.4.2 Interaction of Human and Autonomous Vehicles

The second simulation scenario includes two vehicles, in which three simulation cases are performed by the CarSim vehicle dynamic software. The different cases are distinguished based on the initial position of the intelligent vehicle. In these simulations, the human-driven vehicle is on a primary route, while the intelligent vehicle with predictive cruise control is on a non-primary road, see Fig. 4.7. Note that if only intelligent vehicles are involved it is insufficient to define a hierarchy among the roads, but in the case of mixed traffic scenarios the presence of a human driver requires the rules. Moreover, the safe distance is increased to $s_{safe} = 20$ m to create a sense of confidence for the human participant. It is visualized by the cone in front of the intelligent vehicle.

In the simulations, there is a fault in the operation of the traffic lights; therefore, the proposed speed profile design method must be used. In all scenarios, the predicted time required for the disturbance vehicle to reach the intersection is 1 sec. For safe distance, $s_{safe} = 20$ m is defined.

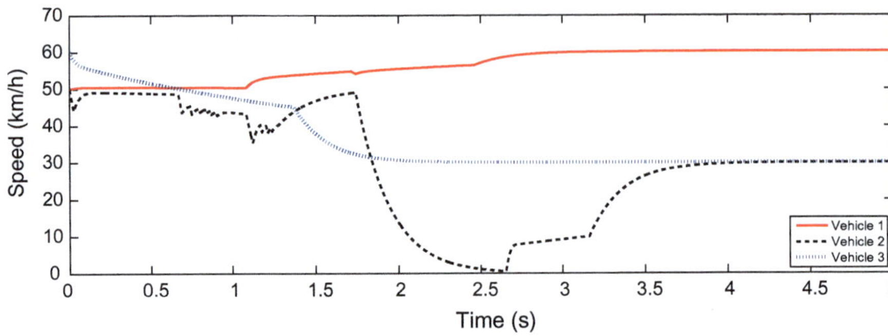

(a) Speed of the vehicles

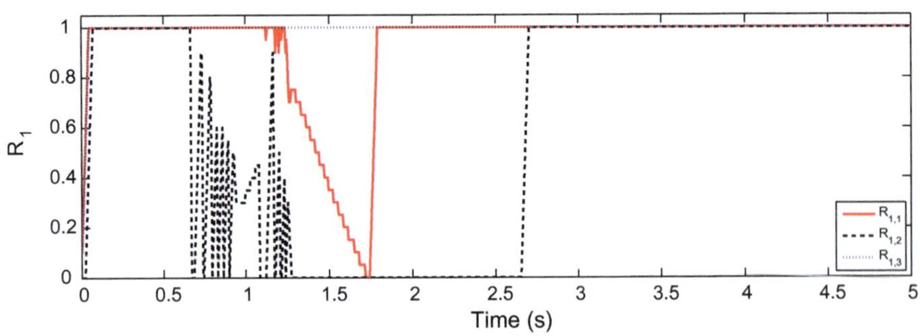

(b) $R_{1,i}$ values

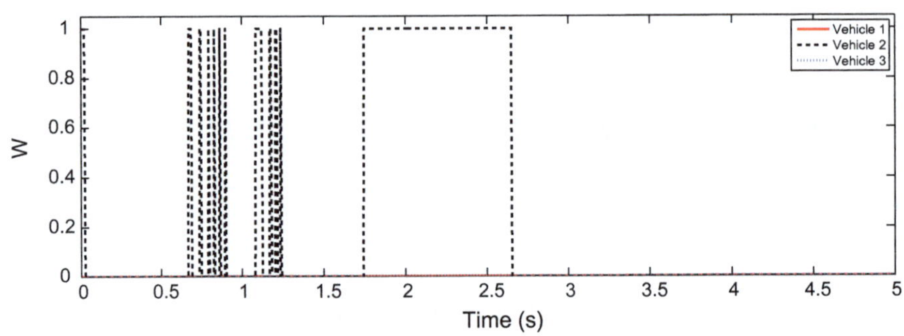

(c) Weight W_i of the vehicles

Fig. 4.6 Results of the simulation with three vehicles

In all simulation cases, the conventional vehicle has the same initial speed and position. Thus, $t_{int} = 8.5$ s in all simulations. The difference between the three cases is the initial position of the controlled vehicle. In Case 1, the intelligent vehicle is initially 90 m from the intersection, while in Case 2 the distance is 110 m, in Case 3 it increased to 150 m. Figure 4.7 shows the time moment t_{int}, when the conventional vehicle reaches the intersection. In Case 1, the cooperation between the vehicles is

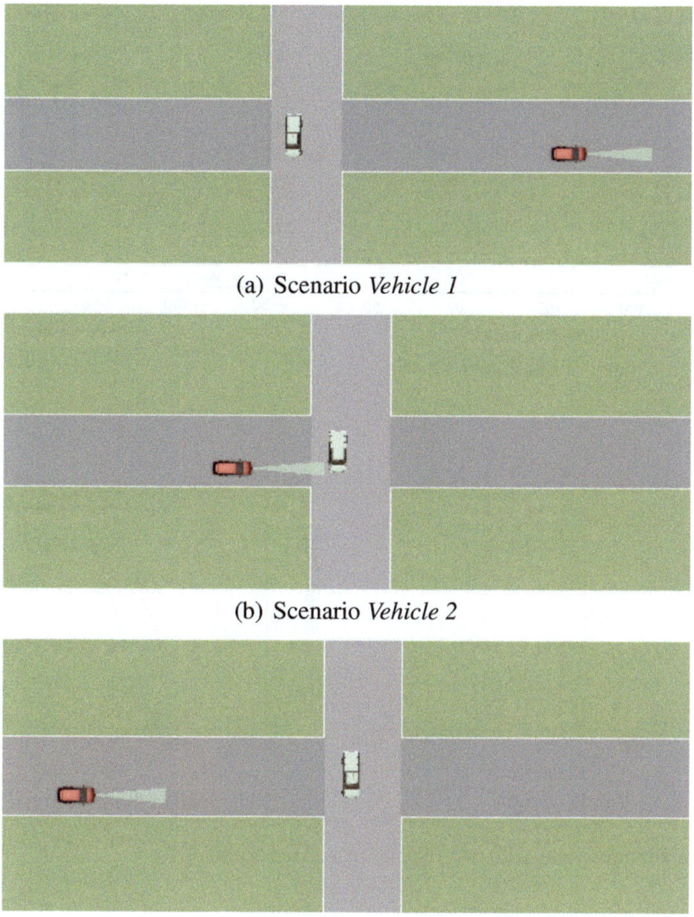

(a) Scenario *Vehicle 1*

(b) Scenario *Vehicle 2*

(c) Scenario *Vehicle 3*

Fig. 4.7 Vehicles in the intersection (intelligent vehicle: red, human-driven vehicle: green)

possible, which means that the intelligent vehicle reaches the intersection first, see Fig. 4.7a. Figure 4.7b illustrates Case 2, when the cooperation is impossible and the speed of the intelligent vehicle must significantly be reduced. However, it can be seen that the conflict in the intersection is avoided and the safe cruising is guaranteed. Case 3 demonstrates the result when the criterion of the maximum set of the cooperation is incorporated in both vehicles. The human-driven vehicle reaches the intersection first and the intelligent vehicle secondly, as illustrated in Fig. 4.7c.

The numerical results of the simulation cases are illustrated in Fig. 4.8. The speeds of the vehicles can be seen in Fig. 4.8a. The speed of the controlled vehicle in Case 1 is increased to reach the intersection rapidly, while in Case 3 the speed is slightly reduced. The reason for that is the selection of different R_1 weights, which

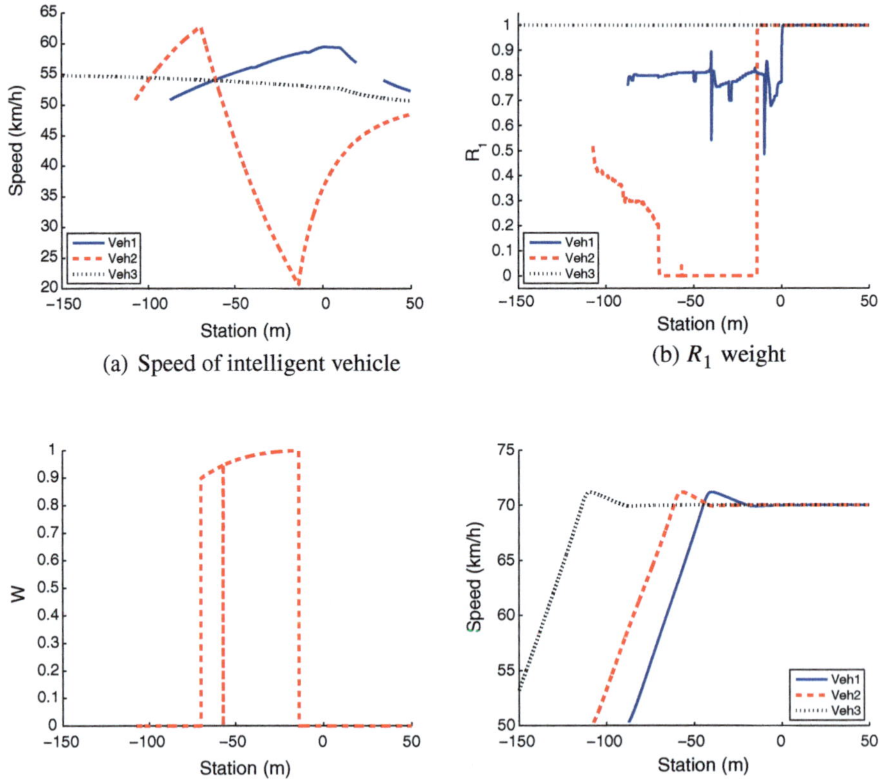

(a) Speed of intelligent vehicle

(b) R_1 weight

(c) W weight

(d) Speed of conventional vehicle

Fig. 4.8 Simulation results with two vehicles in a mixed traffic

is illustrated in Fig. 4.8b. R_1 in Case 3 is continuously at the maximum $R_{1,max} = 1$, while R_1 in Case 1 is slightly reduced. In both cases, the motion of the vehicles can be considered as energy-optimal through the coordination, but in Case 3 the energy consumption is reduced.

In Case 2, the situation in the intersection differs from the previous cases. Until the station point -70 m, the maximum set of coordination contains both vehicles. It is predicted that the intelligent vehicle is able to reach the intersection first, which results in the significant reduction of R_1 and the vehicle speed is increased, see Fig. 4.8a, b. However, after -70 m, the prediction is modified, which means that the coordination cannot be guaranteed. This change is induced by the speed increase of the human-driven vehicle, which is predicted in all computation steps, see Fig. 4.8d. Thus, after the point -70 m, W is activated as depicted in Fig. 4.8c.

References

Aoude G, Desaraju V, Stephens L, How J (2012) Driver behavior classification at intersections and validation on large naturalistic data set. IEEE Trans Intell Transp Syst 13(2):724–736

Asadi B, Vahidi A (2011) Predictive cruise control: utilizing upcoming traffic signal information for improving fuel economy and reducing trip time. IEEE Trans Control Syst Technol 19(3):707–714

Barth A, Franke U (2008) Where will the oncoming vehicle be the next second? In: Proceedings of the IEEE intelligent vehicles symposium, pp 1068–1073

Broadhurst AE, Baker S, Kanade T (2004) A prediction and planning framework for road safety analysis, obstacle avoidance and driver information. In: 11th world congress on intelligent transportation systems, vol 1, pp 1–12

Dang D, Tanwar J, Masood S (2015) A smart traffic solution for high priority vehicles. In: 2015 1st international conference on next generation computing technologies (NGCT), pp 466–470

Dresner K, Stone P (2008) A multiagent approach to autonomous intersection management. J Artif Intell Res 31(1):591–656

Fayazi SA, Vahidi A, Luckow A (2017) Optimal scheduling of autonomous vehicle arrivals at intelligent intersections via MILP. In: 2017 American control conference (ACC), pp 4920–4925

Festag A, Hessler A, Baldessari R, Le L, Zhang W, Westhoff D (2008) Vehicle-to-vehicle and road-side sensor communication for enhanced road safety. In: 15th world congress on intelligent transport systems

Gardner K, Souza C, Hounsell N, Shrestha B, Bretherton D (2009) Review of bus priority at traffic signals around the world. UITP Working Group Technical Report

Hausknecht M, Au TC, Stone P (2011) Autonomous intersection management: multi-intersection optimization. In: 2011 IEEE/RSJ international conference on intelligent robots and systems, pp 4581–4586

Kesting A, Treiber M, Schnhof M, Helbing D (2007) Extending adaptive cruise control to adaptive driving strategies. Trans Res Rec 2000:16–24

Li J, He Q, Zhou H, Guan Y, Dai W (2016) Modeling driver behavior near intersections in hidden markov model. Int J Environ Res Public Health 13(12)

Murgovski N, de Campos GR, Sjöberg J (2015) Convex modeling of conflict resolution at traffic intersections. In: 2015 54th IEEE conference on decision and control (CDC), pp 4708–4713

Nellore K, Hancke G (2016) Traffic management for emergency vehicle priority based on visual sensing. Sensors 16(11) (Basel, Switzerland)

Németh, Csikós A, Gáspár P, Varga I (2013) Design of optimal cruise control considering road and traffic information. In: IFAC joint conference, 5th symposium on system structure and control, Grenoble, France

Riegger L, Carlander M, Lidander N, Murgovski N, Sjöberg J (2016) Centralized MPC for autonomous intersection crossing. In: 2016 IEEE 19th international conference on intelligent transportation systems (ITSC), pp 1372–1377

Smith S, Pavone M, Bullo F, Frazzoli E (2010) Dynamic vehicle routing with priority classes of stochastic demands. SIAM J Control Optim 48(5):3224–3245

Tachet R, Santi P, Sobolevsky S, Reyes-Castro LI, Frazzoli E, Helbing D, Ratti C (2016) Revisiting street intersections using slot-based systems. PloS one 11:e0149607

Wuthishuwong C, Traechtler A (2013) Vehicle to infrastructure based safe trajectory planning for autonomous intersection management. In: 2013 13th international conference on ITS telecommunications (ITST), pp 175–180

Zohdy I, Rakha H (2012) Optimizing driverless vehicles at intersections. In: 19th ITS world congress, Vienna, Austria

Part II
Analysis of the Traffic Flow

Chapter 5
Relationship Between the Traffic Flow and the Cruise Control from the Microscopic Point of View

Introduction and Motivation

The development of energy-efficient strategies for road vehicles has been in the focus of research and development centers in the vehicle industry. One of these methods, which is called the look-ahead control, focuses on the speed control of vehicles. The purpose of the look-ahead control is to design the speed of a vehicle in order to reduce driving energy and fuel consumption while keeping traveling time. Moreover, in the design, the road information such as road slopes and speed limits and the local traffic information such as the current speed, the traffic flow, and the movement of the surrounding vehicles are taken into consideration. The look-ahead control can be considered as an extension of adaptive cruise control with road and traffic information.

A powertrain model was proposed to analyze features such as vehicle power, fuel consumption, acceleration, speed, and position in Rakha et al. (2012). A road type and congestion level estimation method was combined with the principal components analysis for a variety of purposes, e.g., traffic information systems, intelligent real-time control systems, and energy consumption/emissions in Zhu and Barth (2006).

Although the influence of the road and traffic information on the look-ahead control has been presented in several publications, a smaller emphasis has been placed on traffic flow impair caused by speed reduction. There is a relationship between the traffic flow and the look-ahead control; they interact strongly with each other. Thus, in the design the impacts of the traffic flow on the look-ahead control must be taken into consideration. A method is developed which considers the speed of the traffic flow and improves the effect of the look-ahead control on the traffic flow. The effect of the traffic flow on the look-ahead control method will be analyzed in the chapter. Since the speeds of the individual vehicles also affect the speed of the traffic flow, a sensitivity analysis of the parameter variation in the look-ahead control will be performed. If the traffic information is also considered in the look-ahead control, an undesirable side effect on the traffic flow may occur.

P. Gáspár and B. Németh, *Predictive Cruise Control for Road Vehicles Using Road and Traffic Information*, Advances in Industrial Control,
https://doi.org/10.1007/978-3-030-04116-8_5

In the motivation section, the trade-off between the traffic flow and the look-ahead control is illustrated through two simulation examples. The first example shows that the efficient traffic flow requires the information of the traffic flow speed. Thus, in the look-ahead control, the traffic flow speed information must be taken into consideration. At the same time, the look-ahead control may be disadvantageous when all of the vehicles in the traffic flow incorporate the traffic speed information without any further consideration. The second example shows an illustration of this side effect.

The first example proposes the motivation, i.e., why the incorporation of the traffic flow speed is important in the look-ahead control. In the example, a vehicle is shown which is driven along a highway at 130 km/h, see Fig. 5.1a. It illustrates the topography data of a real European highway section, derived from Google Earth. On the forthcoming sections from 1600 m, there is a congestion, in which the vehicles are moving at 90 km/h. The first vehicle, which is equipped with a look-ahead control and traffic information, receives the speed of the forthcoming traffic flow in the look-ahead algorithm, and thus the vehicle speed is reduced gradually within a relatively long time until the vehicles reach the congestion, see the illustration with dashed

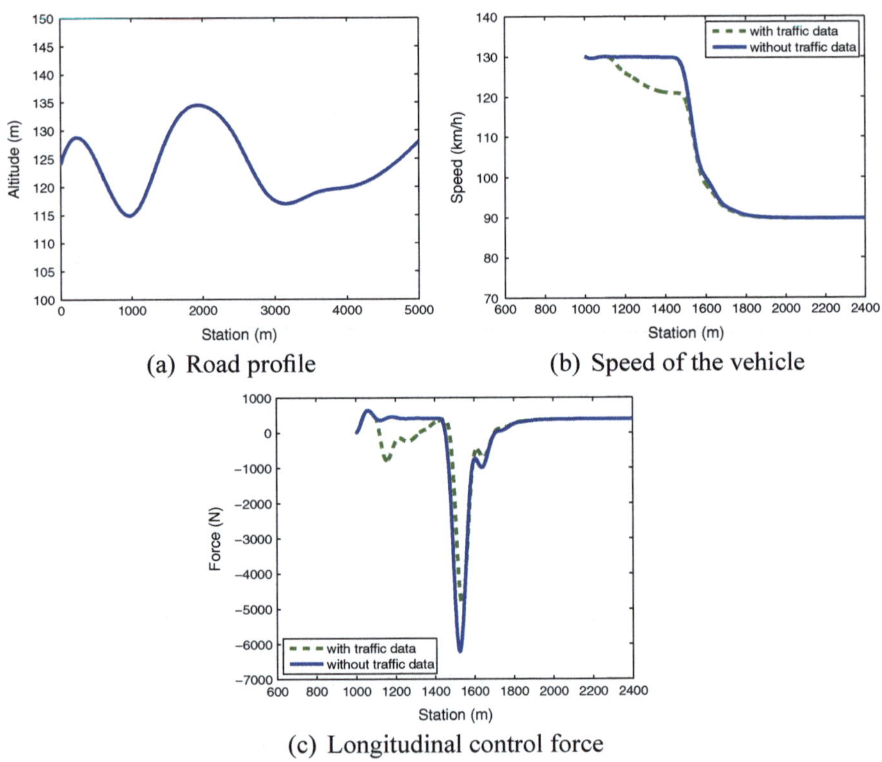

(a) Road profile (b) Speed of the vehicle

(c) Longitudinal control force

Fig. 5.1 Example of the advantage of traffic information consideration green line in Fig. 5.1b. In the case of the second vehicle, the traffic flow information is not considered, and thus the speed of the vehicle must be reduced abruptly when the vehicle arrives at the congestion, see the illustration with the solid blue line. The consideration of the traffic flow information requires smaller but longer longitudinal control force to reduce the velocity of the vehicle, see Fig. 5.1c. The advantage of this solution is to avoid abrupt speed reduction. In the second case, the braking is shorter but significantly more intensive, which is dangerous in the traffic.

In the second example, several vehicles equipped with look-ahcad control and traffic speed information are traveling one after another along the same road. Each vehicle uses the look-ahead control method and calculates its own speed to minimize control energy and fuel consumption. Note that the traffic flow speed is necessarily determined by the speed of the individual vehicles. In the simulation, each vehicle uses road information about 500 m section ahead of the vehicle. Since the distance between two vehicles in the example is 50 m, the 500 m ranges overlap. Figure 5.2 shows the motions of some vehicles in the rush-hour traffic. In the computation of their optimal speeds, the information about the traffic flow speed on the oncoming road section is available. Two scenarios are analyzed in the example: in the first

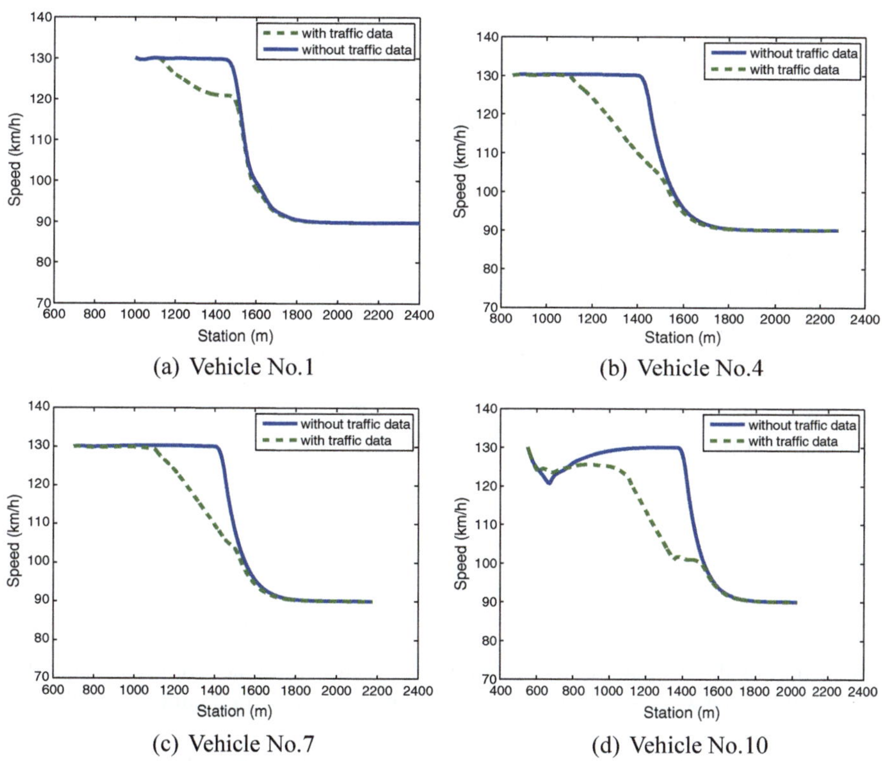

Fig. 5.2 Effect of the look-ahead control on traffic flow

scenario, the vehicles consider the traffic flow information (dashed green lines), while in the second scenario it is not considered (solid blue lines).

The speed profiles of the first vehicles No. 1 are shown in Fig. 5.2a. Since in the first scenario the vehicle has information about the forthcoming congestion, it reduces the speed from section point 1100 m. The speed reduction has an effect not only on the fuel consumption but also on the traffic flow. The energy-optimum speed considering the traffic flow significantly decreases, which is illustrated by dashed line for vehicles No. 4, No. 7, and No. 10 in Figs. 5.2b, c, d. As it is shown, the look-ahead control results in a significant speed reduction of the traffic flow compared to look-ahead control in which the traffic flow is not considered. Moreover, the rate of speed reduction decreases along the flow and the distances increase. Consequently, it is necessary to elaborate an improved look-ahead strategy in order to avoid this significant speed reduction, and a balance between the energy-efficiency and the traffic flow consideration must be analyzed.

5.1 Sensitivity Analysis of the Optimum Solution

In this section, the relation between the original reference speed $v_{ref,i}$ and the modified reference speed λ is analyzed. Since the speed of the traffic flow also influences the reference speed $v_{ref,i}$, its variation has a significant role in the design of reference speed λ. The selection of λ depends on both explicit and implicit factors:

$$\lambda = \mathrm{f}(v_{ref,i}, Q, \gamma_i) = \mathrm{f}(v_{ref,i}, Q(v_{ref,i}), \gamma_i(v_{ref,i})). \tag{5.1}$$

The explicit factors depend on (5.1) through the weighted expression (2.26), while the implicit factors depend on the optimization process (2.41), where the values β_i depend on $v_{ref,i}$.

The reference signal λ is formed in (3.7). In the expression, energy-optimum reference speed $\bar{\lambda}$ is the only part which depends on $v_{ref,i}$. If $v_{ref,i}$ is modified, then the new energy-optimum solution is $\bar{\lambda}^2 + \delta$ in (3.7) instead of $\bar{\lambda}^2$. Here, δ represents the deviation from $\bar{\lambda}^2$, which is caused by the change of $v_{ref,i}$. In this case, the new reference speed λ_{new} is

$$\begin{aligned}\lambda_{new} &= \sqrt{R_1\delta + R_1\bar{\lambda}^2 + v_{ref,0}^2(1 - R_1)} \\ &= \sqrt{R_1\delta + \lambda^2}.\end{aligned} \tag{5.2}$$

The equation shows that the modification in δ occurs in the new reference signal. If R_1 is small, then $\lambda_{new} \to \lambda$. If R_1 increases, then δ has a significant role in the new reference speed λ_{new}.

The relationship between R_1, δ and λ shows that the speed of the vehicle using look-ahead control is determined by the weight R_1. Consequently, this weight also influences the traffic flow through λ. In a high-density traffic, the selection of R_1

has a significant role. If the vehicles in the traffic simultaneously apply high R_1 to reduce energy, even a small modification of $v_{ref,i}$ has a large impact on the speed of the vehicles. Thus, a high R_1 results in the propagation of the deviation δ on the traffic flow.

In the following, the deviation δ is analyzed together with the sensitivity analysis of $\bar{\lambda}$. δ is represented by the total derivative of $\bar{\lambda}^2$, such as

$$\delta = \sum_{i=1}^{n} \frac{\partial(\bar{\lambda}^2)}{\partial v_{ref,i}} dv_{ref,i}, \tag{5.3}$$

where $dv_{ref,i}$ values are the variations of the reference speeds in section point i. For the further analysis, the derivative $\partial(\bar{\lambda}^2)/\partial v_{ref,i}$ is computed:

$$\begin{aligned}\frac{\partial(\bar{\lambda}^2)}{\partial v_{ref,i}} &= \frac{\partial Q}{\partial v_{ref,i}} v_{ref,0}^2 + \sum_{j=1}^{n} \frac{\partial \gamma_j}{\partial v_{ref,i}} v_{ref,j}^2 + 2\gamma_i v_{ref,i} + \\ &+ \frac{2}{m} \sum_{j=1}^{n} s_j F_{dj,r} \sum_{k=j}^{n} \frac{\partial \gamma_k}{\partial v_{ref,i}} + 2 s_1 \frac{\partial Q}{\partial v_{ref,i}} (\ddot{\xi}_0 + g \sin\alpha).\end{aligned} \tag{5.4}$$

This derivative contains several dependence factors concerning $v_{ref,i}$, formulated in partial derivatives. Thus, it is necessary to compute derivatives $\frac{\partial \gamma_k}{\partial v_{ref,i}}$ and $\frac{\partial Q}{\partial v_{ref,i}}$. Since in the computation process Q is gridded and fixed in each section point, the nonlinear optimization process is transformed into a constrained quadratic optimization process. The values $\frac{\partial \gamma_k}{\partial v_{ref,i}}$ are computed from the result of the constrained quadratic optimization problem (2.41). The $v_{ref,i}$ dependence is derived from the Lagrange function. The cost function of the optimization problem is

$$\bar{F}_{l1}^2 = (\beta_0(\bar{Q}) + \beta_1(\bar{Q})\bar{\gamma}_1 + \cdots + \beta_n(\bar{Q})\bar{\gamma}_n)^2, \tag{5.5}$$

where the constraints are $0 \le \gamma_i \le 1$ and $Q + \sum_{j=1}^{n} \gamma_j = 1$. Using the constrains, the Lagrange function is formed as

$$\begin{aligned}\mathscr{L} &= \left(\beta_0(\bar{Q}) + \beta_1(\bar{Q})\bar{\gamma}_1 + \cdots + \beta_n(\bar{Q})\bar{\gamma}_n\right)^2 + \\ &+ \sum_{j=1}^{n} u_j\left(\gamma_j - 1\right) + \sum_{j=n+1}^{2n} u_j\left(-\gamma_j\right) + u_{2n+1}(Q + \sum_{j=1}^{n} \gamma_j - 1),\end{aligned} \tag{5.6}$$

where u_is are the Lagrange multipliers of the system. The sensitivity analysis is represented by the following derivation Fiacco (1983):

$$\frac{\partial \Gamma}{\partial v_{ref,i}} = -\left(\nabla^2 \mathscr{L}\right)^{-1} \cdot \frac{\partial}{\partial v_{ref,i}} \nabla \mathscr{L}, \tag{5.7}$$

where

$$\Gamma = \begin{bmatrix} \gamma_1(v_{ref,i}) \\ \vdots \\ \gamma_n(v_{ref,i}) \\ u_1(v_{ref,i}) \\ \vdots \\ u_{2n+2}(v_{ref,i}) \end{bmatrix} \tag{5.8}$$

and furthermore

$$\nabla \mathcal{L} = \begin{bmatrix} 2\beta_0\beta_1 + 2\beta_1 \sum_{j=1}^{n} \beta_j \gamma_j + u_1 - u_{n+1} + u_{2n+1} \\ \vdots \\ 2\beta_0\beta_n + 2\beta_n \sum_{j=1}^{n} \beta_j \gamma_j + u_n - u_{2n} + u_{2n+1} \\ \gamma_1 - 1 \\ \vdots \\ \gamma_n - 1 \\ -\gamma_1 \\ \vdots \\ -\gamma_n \\ Q + \sum_{j=1}^{n} \gamma_j - 1 \end{bmatrix} \tag{5.9}$$

$$\nabla^2 \mathcal{L} = \begin{bmatrix} \nabla_1 & I_{n\times n} \; -I_{n\times n} \; J_{n\times 1} \\ I_{n\times n} & 0_{n\times 2n+1} \\ -I_{n\times n} & 0_{n\times 2n+1} \\ J_{1\times n} & 0_{1\times 2n+1} \end{bmatrix}, \tag{5.10}$$

where $I_{n\times n}$ represents a square n-dimensional identity matrix, J is the matrix of ones, 0 is the matrix of zeros, and

$$\nabla_1 = \begin{bmatrix} 2\beta_1^2 & \cdots & 2\beta_1\beta_n \\ 2\beta_2\beta_1 & \cdots & 2\beta_2\beta_n \\ \vdots & \vdots & \vdots \\ 2\beta_n\beta_1 & \cdots & 2\beta_n^2 \end{bmatrix} \tag{5.11}$$

is an n-dimensional square matrix of the values β_i. The matrix $\nabla\mathcal{L}$ is transformed as

$$\nabla \mathcal{L} = \sigma + \nabla^2 \mathcal{L} \Gamma, \tag{5.12}$$

where

$$\sigma = \left[2\beta_0\beta_1 \cdots 2\beta_0\beta_n \ -J_{n\times 1} \ 0_{n\times 1} \ Q-1\right]^T. \tag{5.13}$$

Considering (5.12), the relation (5.7) is transformed as

$$\frac{\partial \Gamma}{\partial v_{ref,i}} = -\left(\nabla^2 \mathscr{L}\right)^{-1} \frac{\partial \sigma}{\partial v_{ref,i}} - \left(\nabla^2 \mathscr{L}\right)^{-1} \frac{\partial(\nabla^2 \mathscr{L})}{\partial v_{ref,i}}\Gamma - \left(\nabla^2 \mathscr{L}\right)^{-1} \nabla^2 \mathscr{L} \frac{\partial \Gamma}{\partial v_{ref,i}}. \tag{5.14}$$

Thus, the derivative of Γ is expressed as

$$\frac{\partial \Gamma}{\partial v_{ref,i}} = -\frac{1}{2}\left(\nabla^2 \mathscr{L}\right)^{-1} \left(\frac{\partial \sigma}{\partial v_{ref,i}} + \frac{\partial(\nabla^2 \mathscr{L})}{\partial v_{ref,i}}\Gamma\right). \tag{5.15}$$

Note that the expression (5.15) can be simplified to obtain a more compact solution. First, in term $\partial\sigma/\partial v_{ref,i}$ only the elements $1 \ldots n$ depend on $v_{ref,i}$. Therefore, the elements $n+1 \cdots 3n+1$ are zero. Second, in term $\partial(\nabla^2\mathscr{L})/\partial v_{ref,i}$ also the block ∇_1 depends on $v_{ref,i}$. Therefore, in the multiplication $\partial(\nabla^2\mathscr{L})/\partial v_{ref,i} \cdot \Gamma$ the coefficients of $u_1 \ldots u_{2n+1}$ are zero.

These two statements mean that the expression $\left(\frac{\partial\sigma}{\partial v_{ref,i}} + \frac{\partial(\nabla^2\mathscr{L})}{\partial v_{ref,i}}\Gamma\right)$ is an $3n+1 \times 1$ vector, whose elements $n+1 \cdots 3n+1$ are zero. Thus, it is possible to handle $(\nabla^2\mathscr{L})^{-1}$ as an inverse of a block matrix, and the partition $n \times n$ is sufficient for the further solution

$$\nabla^2 \mathscr{L} = \left[\begin{array}{c|c} \nabla_1 & E \\ \hline E^T & 0_{2n+1\times 2n+1} \end{array}\right]. \tag{5.16}$$

The inverse $(\nabla^2\mathscr{L})^{-1}$ is partitioned as

$$(\nabla^2 \mathscr{L})^{-1} = \left[\begin{array}{c|c} B_{11} & B_{12} \\ \hline B_{21} & B_{22} \end{array}\right]. \tag{5.17}$$

By applying the matrix inversion lemma, the necessary block B_{11} is computed as

$$\begin{aligned} B_{11} &= \nabla_1^{-1} + \nabla_1^{-1} E(-E^T \nabla_1^{-1} E)^{-1} E^T \nabla_1^{-1} \\ &= \left(\nabla_1 + E\left(-2^T \nabla_1^{-1} E\right)^{-1} E^T\right)^{-1} \\ &= \left(\nabla_1 - \frac{1}{2} E E^{-1} \nabla_1 (E^T)^{-1} E^T\right)^{-1} \\ &= 2\nabla_1^{-1}. \end{aligned} \tag{5.18}$$

Finally, the resulting $\frac{\partial \Gamma}{\partial v_{ref,i}}$ (5.7) is the following:

$$\begin{bmatrix} \frac{\gamma_1}{v_{ref,i}} \\ \vdots \\ \frac{\gamma_n}{v_{ref,i}} \end{bmatrix} = -\nabla_1^{-1} \frac{\partial \beta_i}{\partial v_{ref,i}} \begin{bmatrix} 2\beta_1\gamma_i \\ \vdots \\ 2\beta_{i-1}\gamma_i \\ 2\beta_i\gamma_i + 2\beta_0 + \sum_{j=1}^{n} \beta_j\gamma_j \\ 2\beta_{i+1}\gamma_i \\ \vdots \\ 2\beta_n\gamma_i \end{bmatrix}. \tag{5.19}$$

The computed $\gamma_i / v_{ref,i}$ are substituted in (5.4), which results in the required derivative $\partial(\bar{\lambda}^2)/\partial v_{ref,i}$.

Finally, it is noted that the computation of ∇_1^{-1} is a crucial part of the sensitivity analysis. However, ∇_1 is a singular matrix; consequently, the numerical computation of ∇_1^{-1} is hard and almost impossible; moreover, the condition number of ∇_1 is very high. Therefore, in practice, the Moore–Penrose pseudo-inverse ∇_1^{+} is applied in the analysis instead of ∇_1^{-1}.

5.1.1 Example of the Sensitivity Analysis

In the following, an example is shown, which gives an illustration of the analysis based on the derivative $\partial(\bar{\lambda}^2)/\partial v_{ref,i}$. In the analysis, a vehicle with 2000 kg mass is considered, which travels on a flat highway ($\alpha_i = 0$). The vehicle considers information on the horizon 150 m ahead of the vehicle, which results in three section points. There is a speed limit on the road, and thus the reference speeds at the section points are $v_{ref,0} = 130$ km/h, $v_{ref,1} = 130$ km/h, $v_{ref,2} = 90$ km/h, and $v_{ref,3} = 90$ km/h. The current speed of the vehicle is $\dot{\xi}_0 = 125$ km/h and the acceleration is $\ddot{\xi}_0 = 0.1\,\text{m/s}^2$. In the analysis, the effect of $v_{ref,3}$ variation is examined. Thus, the computation of $\partial(\bar{\lambda}^2)/\partial v_{ref,3}$ is required, see (5.4) and (5.19). The elements of (5.19) are

$$\nabla_1 = 10^9 \cdot \begin{bmatrix} 1.3604 & 0.6520 & 0.6520 \\ 0.6520 & 0.3125 & 0.3125 \\ 0.6520 & 0.3125 & 0.3125 \end{bmatrix};$$

$$\frac{\partial \beta_3}{\partial v_{ref,3}} = 1000;$$

$$\begin{bmatrix} 2\beta_1\gamma_3 \\ 2\beta_2\gamma_3 \\ 2\beta_3\gamma_3 + 2\beta_0 + \sum_{j=1}^{3} \beta_j\gamma_j \end{bmatrix} = \begin{bmatrix} 26080.2 \\ 12500 \\ -15253.6 \end{bmatrix}$$

The result of (5.19) is

$$\begin{bmatrix} \frac{\gamma_1}{v_{ref,3}} \\ \frac{\gamma_2}{v_{ref,3}} \\ \frac{\gamma_3}{v_{ref,3}} \end{bmatrix} = \begin{bmatrix} -91144884 \\ -43684898 \\ 233851386 \end{bmatrix},$$

and therefore the resulting derivative from (5.4) is $\partial(\bar{\lambda}^2)/\partial v_{ref,3} = 8.74$. Since in the example only one parameter is modified, i.e., $v_{ref,3}$, the derivation delta resulting from (5.3) is

$$\delta = \frac{\partial(\bar{\lambda}^2)}{\partial v_{ref,3}} dv_{ref,3} = 8.74 \cdot dv_{ref,3}.$$

The modified new reference speed λ_{new} is computed from (5.2) as

$$\lambda_{new} = \sqrt{8.74 \cdot dv_{ref,3} R_1 + \lambda^2},$$

where λ yielded by (5.1).

In the following, the results of the sensitivity analysis as illustrations are presented. Table 5.1 shows scenarios with different R_1 and $dv_{ref,3}$ values. Since the aim of the computation is to show the effect of a traffic congestion on the reference speed, $v_{ref,3}$ value is reduced, and therefore $dv_{ref,3} \leq 0$. In case of $R_1 = 0.25$, the original speed $\lambda = 121.1$ km/h, while at $R_1 = 0.5$ the speed is $\lambda = 111.5$ km/h. The results of the analysis show that the reduction $v_{ref,3}$ induces the decrease of λ_{new}. Moreover, this decrease is also influenced by R_1. If the energy saving of the system has a high priority, then the reduction of λ_{new} compared to λ is higher. Thus, the analysis meets the expectations, see the comments on (5.2). A possible way to balance the reduction in λ_{new} is the decrease of R_1.

Table 5.1 Results of the analysis $dv_{ref,3}$

$R_1 = 0.25$			
$dv_{ref,3}$	−15 km/h	−30 km/h	−45 km/h
λ_{new}	120.6 km/h	120.1 km/h	119.6 km/h
$dv_{ref,3}$	−60 km/h	−75 km/h	−90 km/h
λ_{new}	119.1 km/h	118.7 km/h	118.2 km/h
$R_1 = 0.5$			
$dv_{ref,3}$	−15 km/h	−30 km/h	−45 km/h
λ_{new}	110.5 km/h	109.4 km/h	108.3 km/h
$dv_{ref,3}$	−60 km/h	−75 km/h	−90 km/h
λ_{new}	107.2 km/h	106.1 km/h	104.9 km/h

Table 5.2 Results of the analysis $dv_{ref,2-3}$

$R_1 = 0.25$			
$dv_{ref,2-3}$	−15 km/h	−30 km/h	−45 km/h
λ_{new}	120.1 km/h	119.1 km/h	118.2 km/h
$dv_{ref,2-3}$	−60 km/h	−75 km/h	−90 km/h
λ_{new}	117.1 km/h	116.1 km/h	115.1 km/h
$R_1 = 0.5$			
$dv_{ref,2-3}$	−15 km/h	−30 km/h	−45 km/h
λ_{new}	109.4 km/h	107.2 km/h	105.0 km/h
$dv_{ref,2-3}$	−60 km/h	−75 km/h	−90 km/h
λ_{new}	102.7 km/h	100.4 km/h	98 km/h

The effect of the speed reduction is higher, if both $v_{ref,2}$ and $v_{ref,3}$ are modified. In this case, the δ is computed as

$$\delta = \frac{\partial(\bar{\lambda}^2)}{\partial v_{ref,2}} dv_{ref,2} + \frac{\partial(\bar{\lambda}^2)}{\partial v_{ref,3}} dv_{ref,3} = \\ = 8.73 \cdot dv_{ref,2} + 8.74 \cdot dv_{ref,3}. \tag{5.20}$$

The λ_{new} reference speeds with varying $v_{ref,2}$ and $v_{ref,3}$ are detailed in Table 5.2. The results show that the effect of the multiple reduction in $v_{ref,i}$ is more significant in λ_{new}. In the simulation scenario, $v_{ref,2} = v_{ref,3} \leq 0$.

5.2 Speed Profile Optimization

In the previous sections, the effects of traffic flow speed variation on the look-ahead strategy and the consideration of the traffic flow in the look-ahead are presented. It has been illustrated that the reduction of the parameter variation of $v_{ref,i}$ decreases the reference speed of the vehicle λ_{new}. If the reduction is carried out on several $v_{ref,i}$, their effects on the λ_{new} reduction can be significant. It has a direct effect on the following look-ahead vehicles. Since the speeds of the individual vehicles also affect the speed of the traffic flow, it may lead to an undesirable side effect on the speed of the traffic flow, which is shown in Fig. 5.2. It has also been presented, however, that the impact of parameter variations can be taken into consideration through the performance weight R_1. In this section, an optimization strategy for weight R_1, which handles the individual vehicle energy optimization and its impact on the traffic flow, is elaborated.

Consequently, a further purpose of the speed optimization below is to guarantee that the effect of δ is significantly reduced on the speed design. It practically means that the computed λ_{new} is close to the original λ for all vehicles in the traffic flow.

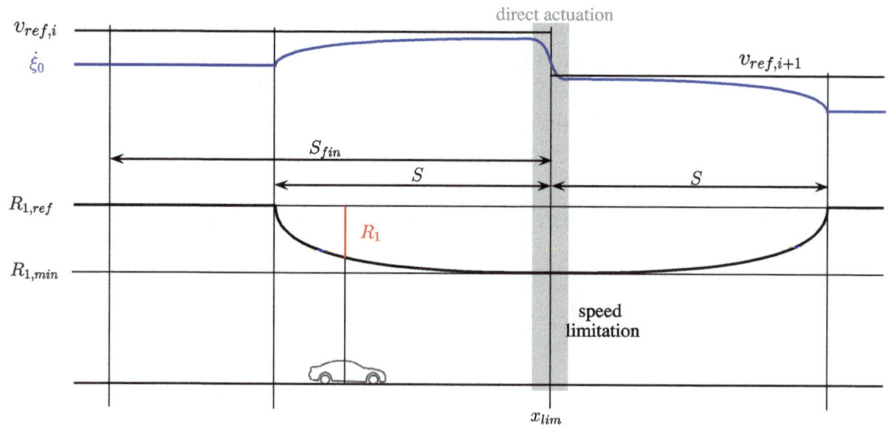

Fig. 5.3 Shape of function R_1

Since R_1 has a significant impact on the effect of δ, it is necessary to construct a function $R_1(x)$, in which x is the position of the vehicle on the road. $R_1(x)$ must provide that $\lambda_{new}(R_1(x)) \rightarrow \lambda(R_{1,ref})$, where the original R_1 is denoted by $R_{1,ref}$.

It is necessary to find a computational method which guarantees the speed of the vehicles in such a way that it tends to the global optimum. An individual decentralized method must be elaborated for the problem of speed limitation. The optimization criterion is defined as follows:

$$\min_{R_1(x)} \sum_{x=-\infty}^{S} |\lambda_{new}(R_1(x)) - \lambda(R_{1,ref})|, \tag{5.21}$$

which considers all of the vehicles on the road section in the interval $x = -\infty \dots S$. Here, S is the section point, where $v_{ref,i}$ is modified due to the congestion. The result of the optimization is an $R_1(x)$ function, which is considered by all the vehicles on the road. Thus, the effect of δ on the speed of the traffic flow speed disappears.

In practice, the shape of $R_1(x)$ is fixed and the number of the optimization parameters is reduced, as illustrated in Fig. 5.3. The chosen shape of the function $R_1(x)$ must guarantee that the speed of the vehicles in the environment of the congestion increases. Thus, before the congestion unnecessary speed reduction can be avoided. However, after the vehicle has reached the congestion speed in x_{lim}, the energy-efficient speed can be applied again. It leads to the increases in weight R_1. The proposed function $R_1(x)$ can be formed as

$$R_1(x) = \begin{cases} R_{1,ref}, & \text{if} x < x_{lim} - S \text{or} \, x_{lim} + S \leq x, \\ R_{1,ref} - (R_{1,ref} - R_{1,min}) \left(\frac{x+S-x_{lim}}{S}\right)^{\varepsilon}, \\ \quad \text{if} \;\; x_{lim} - S \leq x < x_{lim} \\ R_{1,ref} - (R_{1,ref} - R_{1,min}) \left(\frac{x_{lim}+S-x}{S}\right)^{\varepsilon}, \\ \quad \text{if} \;\; x_{lim} \leq x < x_{lim} + S \end{cases} \tag{5.22}$$

where x_{lim} is the position of the congestion. $R_{1,ref}$ is the former reference value R_1, which is valid outside the section $[x_{lim} - S, x_{lim} + S]$. $R_{1,min}$ is the minimum value of R_1 inside the region $[x_{lim} - S, x_{lim} + S]$. ε represents the shape of the function $R_1(x)$ within the interval. In the optimization of $R_{1,ref}$, the values S and x_{lim} are fixed, while $R_{1,min}$ and ε are design parameters. An illustration of $R_1(x)$ is shown in Fig. 5.3.

Therefore, the optimization criterion (5.21) is transformed into

$$\min_{R_{1,min},\varepsilon} \sum_{x=S_{fin}}^{S} |\lambda_{new}(R_{1,min},\varepsilon) - \lambda(R_{1,ref})|, \tag{5.23}$$

where S_{fin} represents the road position where the λ_{new} of the vehicles are considered. Note that the optimization method (5.23) requires knowledge of the current position and speed of each vehicle in the traffic flow. A possible solution of the procedure is the simulation-based optimization, in which the scenario is performed with a given ε and $R_{1,min}$, and then the optimization parameters are tuned to guarantee the criterion (5.23).

5.3 Demonstration of the Optimization Method

In this section, the results of the $R_1(x)$ function optimization are demonstrated through a simulation example, which has already been proposed in section "Introduction and Motivation". In the analysis, two vehicles are compared. The first vehicle is equipped with a look-ahead control and traffic flow information, while the second one uses the look-ahead control but the traffic information is not considered. The aim of the optimization procedure is to reduce the difference between their speeds and calculate the optimum speed profiles.

In the proposed example, the following constant design parameters are chosen: $R_{1,ref} = 0.25$, $S = 500$ m, and $S_{fin} = 1200$ m. The results of the optimization (5.23) are $\varepsilon = 0.1982$ and $R_{1,min} = 0.05$. The look-ahead horizon is chosen as 500 m with 10 sampling points. Figure 5.4 shows the modified speeds of the vehicles in the traffic. As the figures show, the speeds of the vehicles are generally higher than they are in the original scenario, see Fig. 5.2.

Figure 5.5 demonstrates clearly the effect of the R_1 variation on vehicle No. 1. When R_1 is increased, the longitudinal control force and the energy requirement decrease and when R_1 is reduced they increase because terrain characteristics are considered to a lesser extent. At 1100 m, the value R_1 is reduced. Consequently, the speed of the first vehicle increases, because $v_{ref,0} > \bar{\lambda}$. It has an effect on the traffic flow data transmitted to the follower vehicles. After the road section point 2100 m, the performance weight R_1 is set to $R_1 = R_{1,ref} = 0.25$, as a result of which the vehicle speed profile converges to the other scenario, in which traffic data are not considered and $R_1 = R_{1,ref}$.

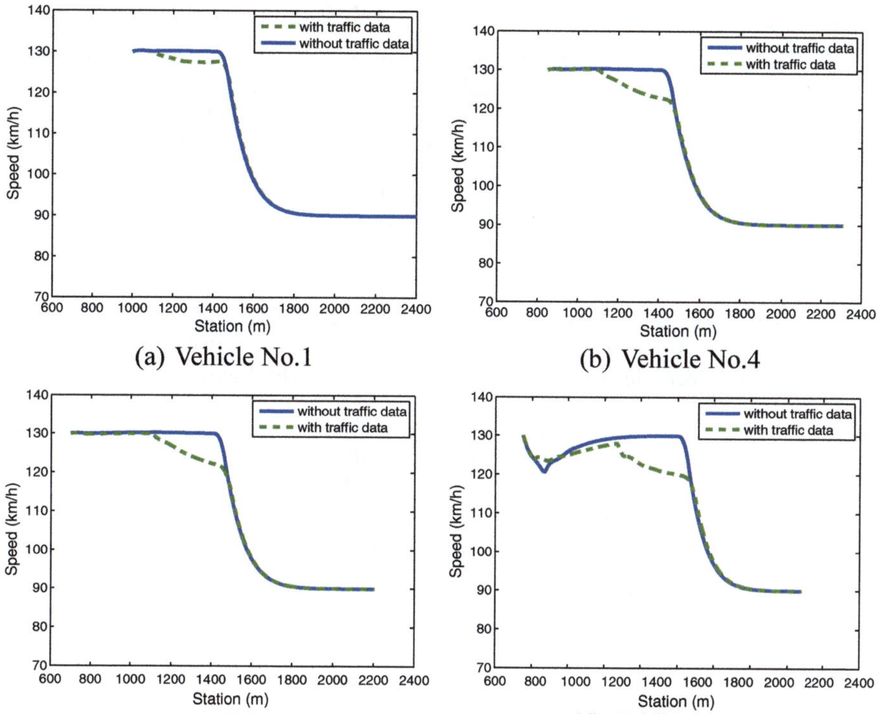

(a) Vehicle No.1 (b) Vehicle No.4

(c) Vehicle No.7 (d) Vehicle No.10

Fig. 5.4 Speed profiles of the vehicles

Fig. 5.5 R_1 of vehicle No. 1

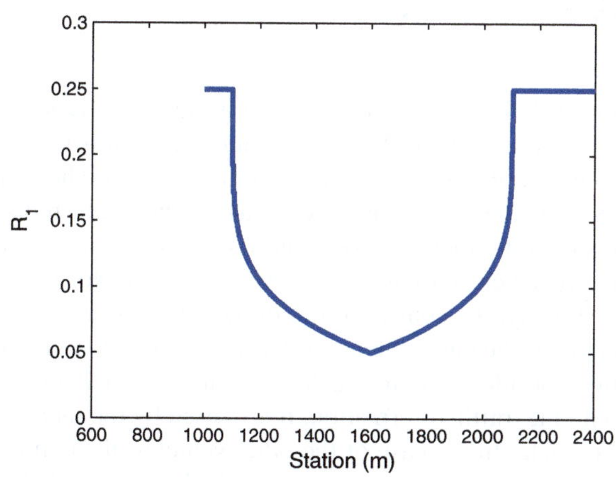

The increase in the speed of vehicle No. 1. has an influence on the speed of the traffic, and thus $v_{ref,i}$ values for the preceding vehicles are higher. As a result, all of the preceding vehicles are faster during the simulation, compared to the example in

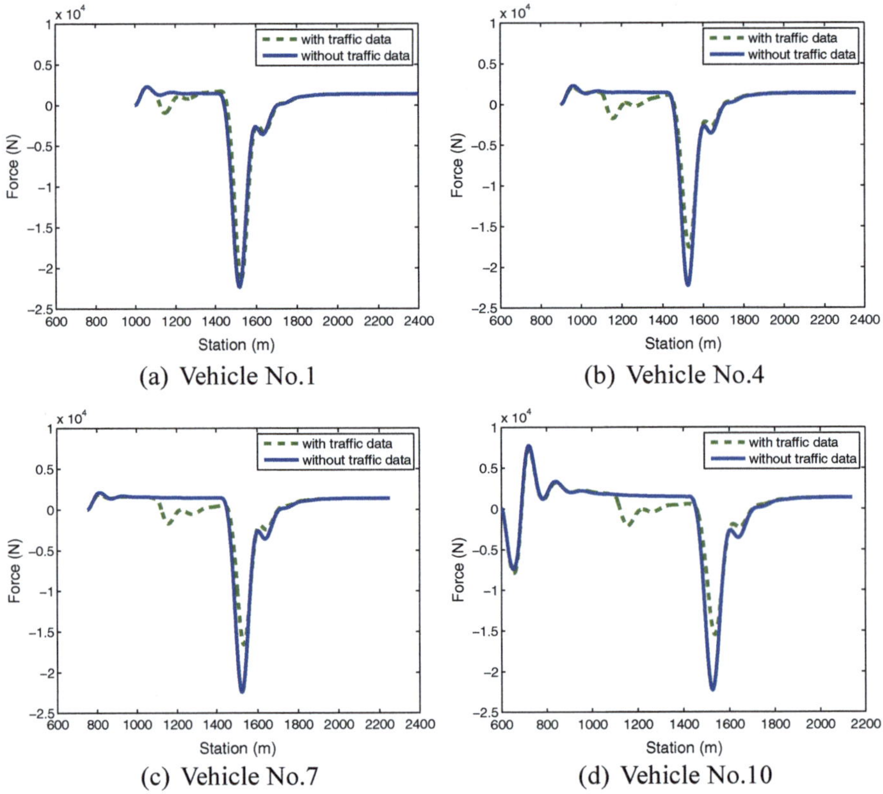

Fig. 5.6 Control forces of the vehicles

Fig. 5.2. The speed modification is significantly smaller, e.g., at 1500 m the speed decrease of vehicle No. 9 is 30 km/h, while it is reduced to 10 km/h in the proposed solution, see Figs. 5.2d and 5.4d. It means that the consideration of traffic flow data requires the reduction in R_1 to avoid the speed reduction of the traffic flow. As a result of the modified look-ahead control, the motion of vehicle No. 9 (Fig. 5.4d) is improved close to the speed level of vehicle No. 1., see Fig. 5.2a.

Finally, the control forces of the vehicles are shown in Fig. 5.6. The illustration contains the data of the look-ahead controlled vehicles with/without traffic information consideration. In Fig. 5.6a, it can be seen that the first vehicle actuates more F_{l1} with the traffic information. It is caused by the increased speed values, compared to Fig. 5.4a. In the case of the other vehicles, the control force is reduced. Generally, the traction energy saving of the vehicles is around 7% compared to the conventional look-ahead solution.

The example has shown that the unfavorable effect of the traffic flow consideration in the look-ahead control can be reduced through the proposed method. The modification of R_1 is effective, and it has a high impact on the speed design. Although the advantage of the traffic flow consideration in the look-ahead control is utilized, the traction energy of the vehicles is reduced.

References

Fiacco AV (1983) Introduction to sensitivity and stability analysis in nonlinear programming, mathematics in science and engineering, vol 165. Academic Press

Rakha H, Kyoungho A, Faris W, Moran K (2012) Simple vehicle powertrain model for modeling intelligent vehicle applications. IEEE Trans Intell Transp Syst 13(2):770–780

Zhu W, Barth M (2006) Vehicle trajectory-based road type and congestion recognition using wavelet analysis. In: IEEE Intelligent transportation systems conference, Toronto, Canada, pp 879–884

Chapter 6
Relationship Between the Traffic Flow and the Cruise Control from the Macroscopic Point of View

Introduction and Motivation

The modeling and analysis of the mixed traffic, in which automated controlled vehicles and conventional uncontrolled vehicles are traveling together, require novel methodologies. The traffic flow model, which was characterized by the automated vehicles, was proposed by Li and Ioannou (2004). The analysis of the traffic flow, in which semi-automated and automated vehicles were traveling together with conventional vehicles were proposed by Schakel et al. (2010); van Arem et al. (2006). Since the semi-automated vehicles supported a smooth traffic flow through their filtering effects, a control law was proposed by Bose and Ioannou (2003). Similar results in the field of mixed traffic were achieved by Zhang and Vahidi (2011). The cooperative cruise control of multiple cars in mixed traffic, in which the motion of the conventional uncontrolled vehicles was considered as the continuous approximation of hybrid dynamical systems, was examined by Chin et al. (2015). This paper illustrated that the cooperative control was able to improve the overall stability. Similarly, the connection of the cruise controlled vehicles was able to improve the string stability through the prevention of shockwave formation and propagation as presented by Talebpour and Mahmassani (2015). The mixed traffic was also modeled and examined from the aspect of the vehicle clusters. A method which was able to predict the distribution of the small clusters in the mixed traffic was proposed by Jerath et al. (2015). Based on this information, the effect of the cruise control vehicles on the mixed traffic was examined. However, in several cases, the traffic flow signals were not measured directly, see e.g., Jia et al. (2016). A method for the estimation of the total density and flow of vehicles for mixed traffic based on the Kalman filtering was proposed by Bekiaris-Liberis et al. (2016). The results were also used for the analysis and control of the vehicles in the traffic.

Although these methods explore complex relationships between the vehicle cruise control and the traffic flow, they were not specified to the predictive cruise control strategy. A method which is specially elaborated for cruise control vehicles was

P. Gáspár and B. Németh, *Predictive Cruise Control for Road Vehicles Using Road and Traffic Information*, Advances in Industrial Control,
https://doi.org/10.1007/978-3-030-04116-8_6

presented by Johannesson et al. (2015). The incorporation of the predictive information about the surrounding vehicles into the convex optimization problem was formulated. As a limitation, in the method, the lane changing of the truck was not considered in the prediction. However, it had a significant impact on the cruising of the connected and automated vehicles as summarized in Bevly et al. (2016). A decision algorithm for predictive control vehicles by which the lane-change maneuvers are based on the consideration of surrounding vehicles was proposed by Gáspár and Németh (2015).

This chapter analyzes the impact of vehicles applying predictive control strategy on the traffic flow. Since the speed profile of the predictive cruise control may differ from that of the conventional vehicle, the characteristics of the traffic flow change. A novel method is applied in which three main parameters concerning the cruise control and the traffic flow are taken into consideration. They are the current traffic density on the road, the ratio of vehicles using cruise control in the traffic, and the energy-efficiency scaling parameter of the cruise control itself. Based on the analysis, an extension of the cruise control method, which considers information of the mixed traffic, will be proposed. The purpose is to create a balance between the traffic flow and energy-efficient traveling. In the selection strategy, the information about the increase in the traffic volume and the ratio of the predictive control vehicles are involved. The method guarantees an energy-efficient motion of the vehicles in the traffic, while an extreme reduction of the traffic speed can be avoided.

6.1 Dynamics of the Traffic with Multi-class Vehicles

The class of the vehicles has a high impact on the average speed and the energy requirements of the entire traffic. In the case of the look-ahead vehicles, the selection of the current speed depends on several factors such as the forthcoming terrain characteristics, road conditions, and speed limits. However, the conventional vehicles are traveling with the maximum speed, if it is possible. It leads to a mixed traffic flow, in which look-ahead controlled vehicles with a speed control are driven together with conventional vehicles. Since the speed profile of the look-ahead control may differ from that of the conventional vehicle, the structure of the traffic flow changes.

There are several methods, in which the modeling of the multi-class traffic is presented, see the survey Wageningen-Kessels (2013). In the following, the method of Ngoduy and Liu (2007) is transformed to the existing traffic modeling problem.

The density-speed relation of the model is a piecewise function, which describes the traffic speed v_u of the different vehicle classes. It has two sections as described below.

- At low traffic density $0 \leq \tilde{\rho} \leq \tilde{\rho}_{crit}$, a linear function approximates the traffic speed, where $\tilde{\rho}_{crit}$ is the critical density. Under this value, the traffic dynamics is handled as a stable system. In this case, free flow is considered, such as

$$v_u = v_{u,max} - \frac{v_{u,max} - v_{crit}}{\tilde{\rho}_{crit}} \tilde{\rho} \tag{6.1}$$

where v_u is the traffic speed of each vehicle classes. Simultaneously, $v_{u,max}$ is the maximum speed of the vehicle class, limited by the speed regulations at free flow.

- At high traffic density $\tilde{\rho}_{crit} \leq \tilde{\rho} \leq \tilde{\rho}_{jam}$ the relationship is nonlinear, where $\tilde{\rho}_{jam}$ is the maximum traffic density in the model. In this case congestion is considered, which is related to an unstable traffic dynamics. The traffic speed is formed as

$$v_u = w \left(\frac{\tilde{\rho}_{jam}}{\tilde{\rho}} - 1 \right) \tag{6.2}$$

where

$$w = \frac{\tilde{\rho}_{crit} v_{crit}}{\tilde{\rho}_{jam} - \tilde{\rho}_{crit}}. \tag{6.3}$$

The parameter v_{crit} is the critical speed of the traffic flow, which is related to $\tilde{\rho}_{crit}$.

In the formulation of the traffic density of the overall system is formed as the unweighted sum of the densities at each traffic flow

$$\tilde{\rho} = \sum_u \rho_u \tag{6.4}$$

In the traffic model, the critical traffic density and the maximum traffic density are computed based on the parameters of each vehicle classes.

It results in the following relations:

$$\tilde{\rho}_{crit} = \rho_{crit} \sum_u \frac{\alpha_u}{\eta_u}, \tag{6.5}$$

$$\tilde{\rho}_{jam} = \rho_{jam} \sum_u \frac{\alpha_u}{\eta_u}, \tag{6.6}$$

where η_u is the passenger-car-equivalent number and α_u is the road fraction parameter. It represents the ratio of each classes from the overall traffic

$$\alpha_u = \frac{\rho_u}{\tilde{\rho}}. \tag{6.7}$$

Figure 6.1 illustrates the function of the density-speed relation. Depending on ρ, it has two different sections, i.e., a linear and a nonlinear one. In this way, the model approximates the dynamics of the traffic appropriately, see Wageningen-Kessels (2013).

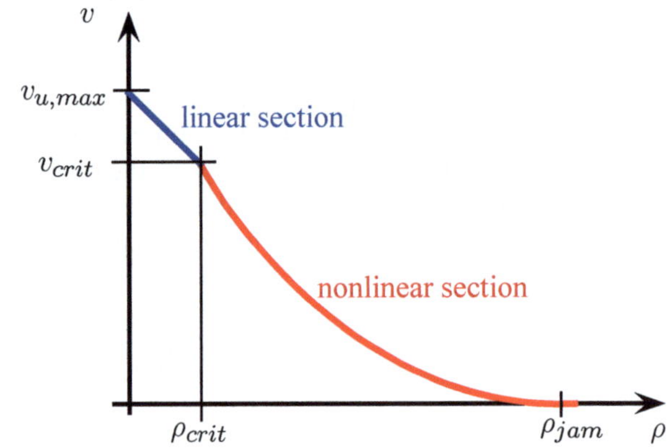

Fig. 6.1 Illustration of the density-speed relation

6.2 Analysis of the Predictive Cruise Control in the Traffic

In the research, the traffic model and the measurements of a test network are built in the VisSim microscopic traffic simulation system. In the demonstration example, a 20 km-long 3-lane segment of the Hungarian M1 highway between Budapest and Tatabnya is modeled in VisSim, in which the terrain characteristics (see Fig. 6.2) and the speed limits are taken into consideration. The speed limit on the section is 130 km/h, although there is a 90 km/h limitation between 5.6 km … 8.5 km segments. Using this model, several simulations with different traffic densities are performed. Moreover, the traffic contains a significant number of look-ahead vehicles, in which the optimization method have been built in.

During the research, several traffic simulations are performed to analyze the impact of the look-ahead controlled vehicles on the entire traffic. In these simulations, the effects of three parameters are examined through their variations, such as

- q_{in}: the inflow of the vehicles on the highway section,

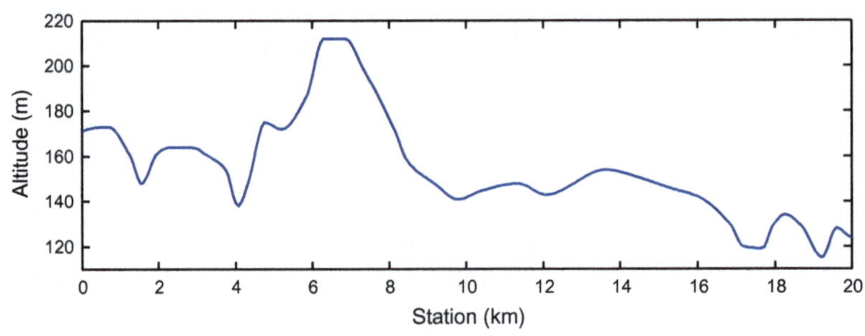

Fig. 6.2 Terrain characteristics of the highway section

Table 6.1 Results of the analyses

q_{in}	$\kappa(\%)$	$R_{1,max}$	Mean of force (N)			Traveling time (s)	Average speed (km/h)
			Conventional	Look-ahead	Overall		
3000	1	0.7	678.1	622.1	677.5	608.3	118.4
3000	20	0.7	676.2	629.4	666.8	609.9	118.1
3000	50	0.7	673.4	630.4	651.9	610.0	118.0
5000	20	0.7	662.8	623.8	566.1	627.3	114.8
3000	20	0.9	674.8	598.7	659.6	613.5	117.4
3000	50	0.9	667.4	598.9	633.2	622.4	115.7

- κ: the ratio of the look-ahead controlled vehicles in the traffic,
- $R_{1,max}$: the upper limit of the energy-efficient parameter in the speed profile optimization.

In the following, the most representative scenarios are presented. The scenarios and the impacts of the controlled vehicles are interpreted, while the numerical results are summarized in Table 6.1. The aims of the analysis are to examine the average speed and the traction forces of the vehicles depending on q_{in}, κ, $R_{1,max}$.

The first scenario is proposed in Fig. 6.3. In this case, the selected parameters are $q_{in} = 3000$ veh/h, $R_{1,max} = 0.7$, and $\kappa = 1\%$. It means that a low number of look-ahead controlled vehicles are on the highway. Consequently, the impact of the automated vehicles on the entire traffic flow is negligible. Thus, the average speed of the traffic is close to its maximum 130 km/h, see Fig. 6.3a. However, in the outer lane the speed is slightly reduced, because the slowest cruise controlled vehicles are all in this lane. It means that a small κ value has a very slight impact on the traffic flow. The average traction forces of the conventional and the cruise controlled vehicles are illustrated in Fig. 6.3b. Since the motions of the look-ahead controlled vehicles are not disturbed by each other or the conventional vehicles, they can achieve their optimal speed. This optimal speed results in force reduction on the entire route, which means that the average traction force is decreased by 8.6%.

In the second scenario, the ratio of the look-ahead vehicles is increased to $\kappa = 20\%$. In this case, the look-ahead vehicles have a high impact on the traffic flow. Figure 6.4a shows that the average speed in the lanes significantly varies, see e.g., in the section between 13 and 20 km. As a result of the increased κ, not all of the look-ahead vehicles are able to realize their speed trajectory. Furthermore, the speed profiles of the conventional vehicles are also influenced by the look-ahead vehicles. The effect of the higher κ on the traction forces is illustrated in Fig. 6.4b. It can be seen that the traction forces of the look-ahead vehicles are closer to those of the conventional vehicles. Thus, due to the increased traffic, not all of the look-ahead vehicles are able to guarantee the fuel economy motion. Furthermore, the traction force of the conventional vehicles slightly decreases. Thus, in this scenario, the look-

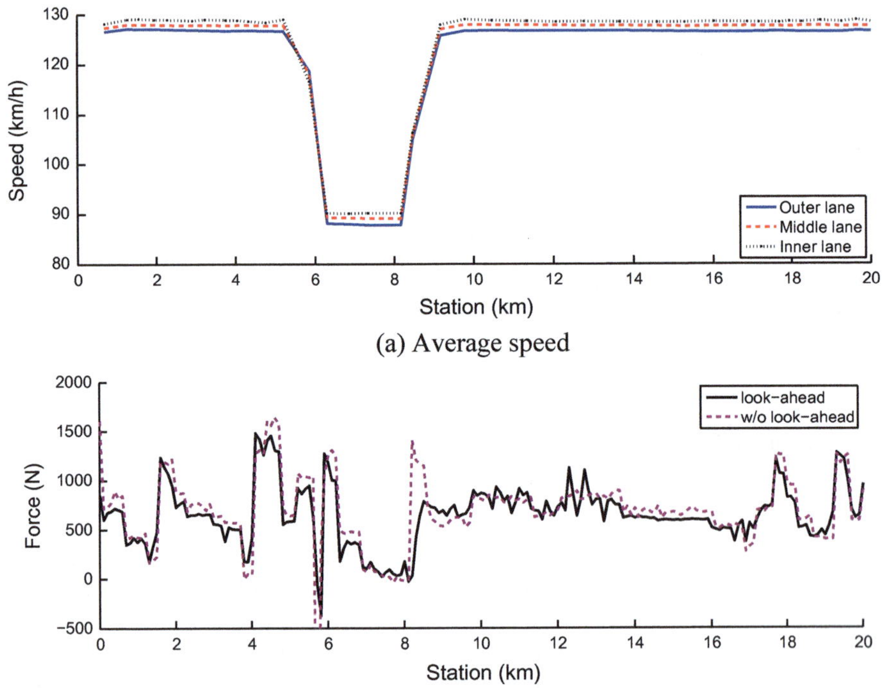

(a) Average speed

(b) Traction force

Fig. 6.3 Scenario 1: $q_{in} = 3000\ veh/h$, $\kappa = 1\%$, $R_{1,max} = 0.7$

ahead vehicles have a low impact on the traffic flow: the speed profile and the traction force of the vehicles without look-ahead control are not modified significantly.

In the third simulation scenario, κ is increased to 50%, which has a high impact on the traffic flow. The results of the scenario are presented in Fig. 6.5. The look-ahead speed profile is forced upon the conventional vehicles in all lanes, see Fig. 6.5a. As a result, the average traction force in the traffic flow decreases, as shown in Fig. 6.5b. This scenario illustrates that the energy consumption of the traffic flow can be influenced by using a high κ ratio. Comparing the results to the first scenario, which contains 99% conventional vehicles, the average traction force of the conventional vehicles decreases from 678.1 N to 673.4 N.

In the fourth scenario, the impact of q_{in} on the traffic flow is illustrated. It means that the inflow increases to $q_{in} = 5000\ veh/h$ value, which represents rush hour traffic. Figure 6.6 presents the results of this scenario. The result of the rush hour traffic is the adaptation of the look-ahead and the conventional vehicles to each other. Since the motion of the look-ahead vehicles is closer to the conventional vehicles, it leads to a slight increase in their traction force, compared to the similar scenario with $q_{in} = 3000\ veh/h$. Moreover, the motion of the conventional vehicles also varies, which results in their force reduction by 2.3% compared to $q_{in} = 3000\ veh/h$. Thus,

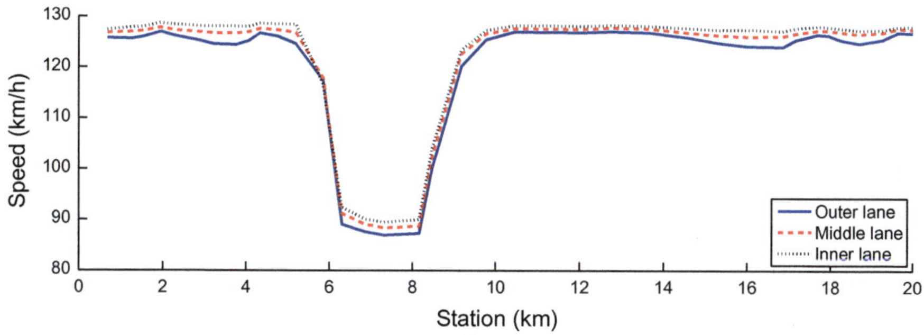

(a) Average speed

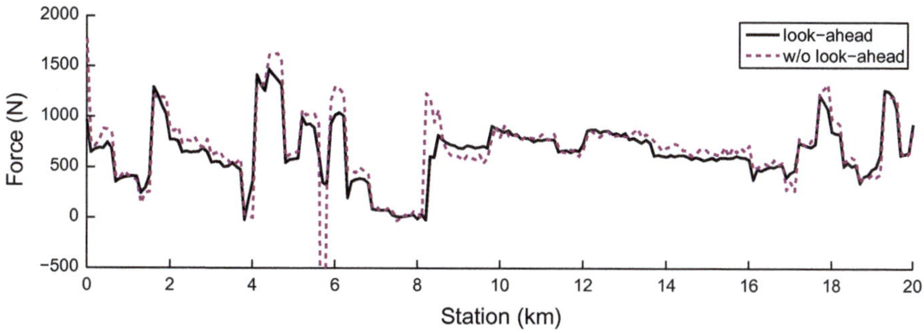

(b) Traction force

Fig. 6.4 Scenario 2: $q_{in} = 3000\ veh/h$, $\kappa = 20\%$, $R_{1,max} = 0.7$

in heavier traffic, the look-ahead control has a significant impact on all the vehicles in terms of force requirement.

The last scenario presents the impact of $R_{1,max}$ increase on the traffic speed and the average traction force, see Figs. 6.7 and 6.8. Due to the increased optimization parameter, the energy-efficiency in the motion of the automated vehicles has a priority. As a consequence, the average speed of the vehicles in the outer lane significantly decreases, compared to the results of the second and third scenarios, see Figs. 6.7a and 6.8a. However, in the other lanes, the average speed is close to the scenario with the parameter $R_{1,max} = 0.7$. Thus, most of the automated vehicles move into the outer lane. As a consequence of these scenarios, the traction forces of the look-ahead vehicles are smaller, and they have little impact on the traction force of the conventional vehicles, as presented in the numerical values of Table 6.1. Thus, the increase in $R_{1,max}$ results in a low energy consumption reduction for the entire traffic.

The force, traveling time and average speed requirements are summarized in Table 6.1. It shows the mean forces of the conventional and the look-ahead vehicles, and their differences are also presented. The following conclusions on the traction forces based on the simulations are drawn:

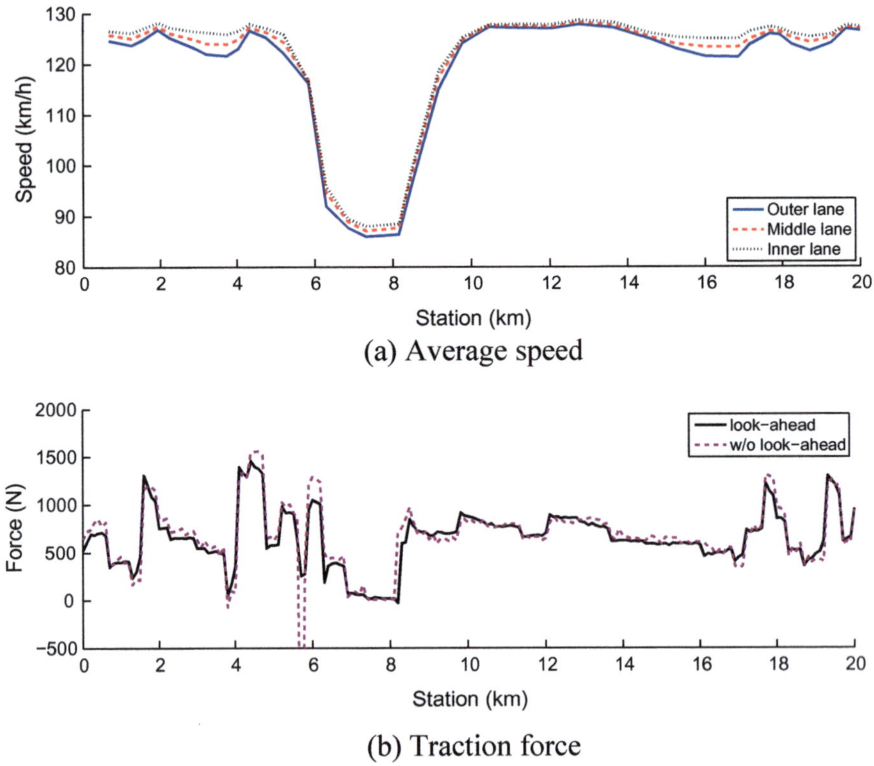

Fig. 6.5 Scenario 3: $q_{in} = 3000\ veh/h$, $\kappa = 50\%$, $R_1 = 0.7$

- The increase in the traffic flow q_{in} can lead to the reduction of the traction forces of the conventional and the look-ahead vehicles simultaneously. However, it increases the traveling time of the vehicles due to the dense traffic.
- If the ratio of the look-ahead vehicles increases, it is slightly disadvantageous for the motion of the look-ahead vehicles, but it improves the force reduction of the conventional vehicles.
- The increase in the $R_{1,max}$ leads to the significant reduction of the look-ahead vehicle forces, and it has high impact on the conventional vehicles.

6.3 Improvement of Traffic Flow Using the Predictive Control

In the previous section, the effects of three parameters on the traffic flow have been examined. It has been demonstrated that the ratio κ of the predictive control vehicles in the traffic flow has significant impact on the energy consumption of all the

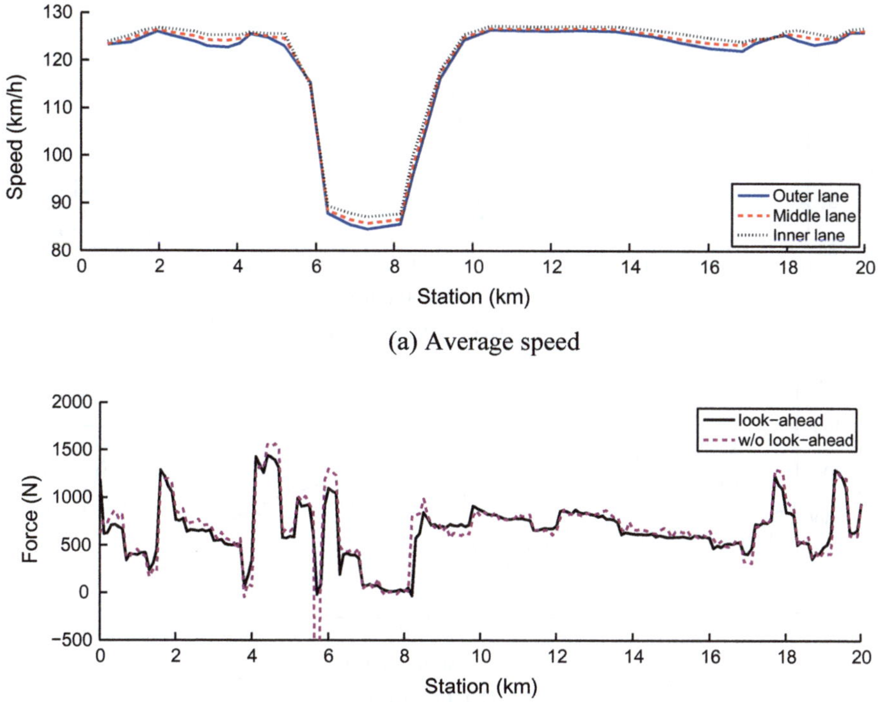

(a) Average speed

(b) Traction force

Fig. 6.6 Scenario 4: $q_{in} = 5000\ veh/h$, $\kappa = 20\%$, $R_{1,max} = 0.7$

vehicles. The weight R_1 reduces the traction force of the predictive control vehicles substantially, and modifies the average speed of the traffic flow. Moreover, the increase of the traffic volume q_{in} results in the reduction of the traffic speed and the traction force simultaneously.

Since the parameters q_{in}, κ and R_1 have an impact on the traffic speed and the traction force, an appropriate speed profile design for the predictive control vehicles must be found. In a real traffic scenario, the measurement of q_{in} and the information about κ are available. Furthermore, the weight R_1 in the predictive control vehicles is a parameter by which the entire traffic can be influenced. With an appropriate choice of R_1, the energy efficiency and the average speed of the traffic flow can be improved. In the following, a strategy for the selection of R_1 is proposed, by which the effect of the predictive control on the traffic flow is considered.

The core of the strategy is based on the design of $R_{1,max}$. This parameter has a relevance in the design of R_1, see (3.14). It means that $R_{1,max}$ determines the maximum value of R_1. In the optimization, (3.14) $R_{1,max}$ is an upper bound of the optimization variable R_1. Since the weight R_1 has a high impact on the traffic flow (see Fig. 6.7), the bound $R_{1,max}$ is also important in the limitation of the traffic flow speed reduction.

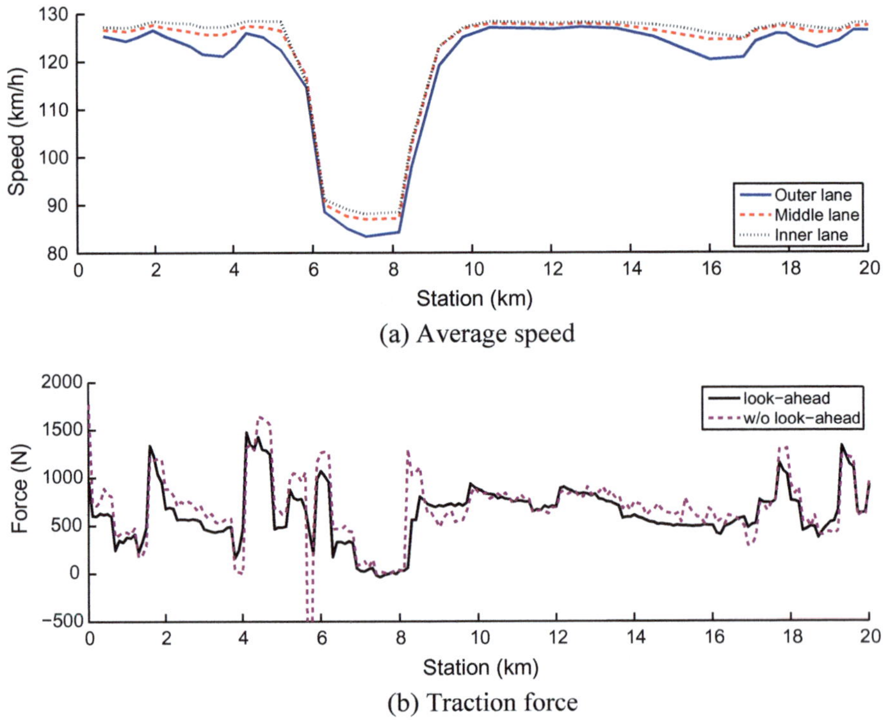

Fig. 6.7 Scenario 5: $q_{in} = 3000\ veh/h$, $\kappa = 20\%$, $R_{1,max} = 0.9$

In the $R_{1,max}$ selection strategy the information about q_{in} and κ are involved.

- Through the appropriate selection of $R_{1,max}$ the disadvantageous effect of the high q_{in} on the traffic speed can be reduced, see Fig. 6.6. If the R_1 weight is reduced in rush hour traffic, the speed profile of the predictive control vehicle is closer to that of the conventional vehicle. As a result, the adaptation of the predictive control vehicles to the conventional vehicles leads to an increase in the traffic speed.
- The ratio κ has a high impact on the traffic, see Fig. 6.5. If κ is high, the average speed of the entire traffic decreases. Although it results in an energy saving traffic flow, the motion of the conventional vehicle can be significantly inhibited. Therefore, it is recommended to limit $R_{1,max}$ to avoid the disadvantageous effect of the high κ values.

The assumptions of q_{in} and κ are formulated in a function, such as

$$R_{1,max} = \mathrm{f}(q_{in}, \kappa) \tag{6.8}$$

where f is an appropriately chosen function that may depend on the current road section and the traffic requirements. For example, there are some road sections where the fast motion of the vehicles is more important than saving energy. In this case, f

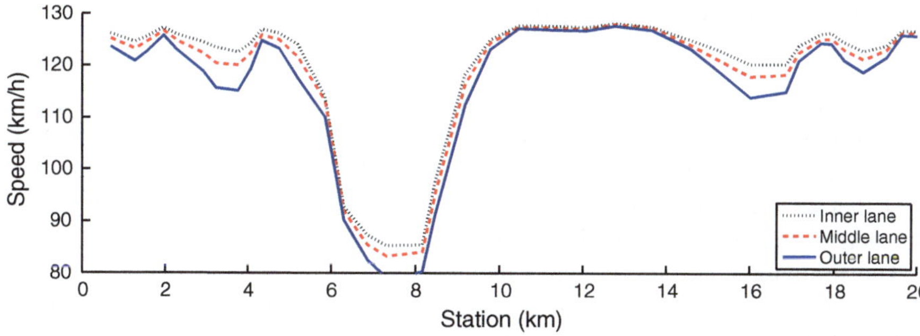

(a) Average speed

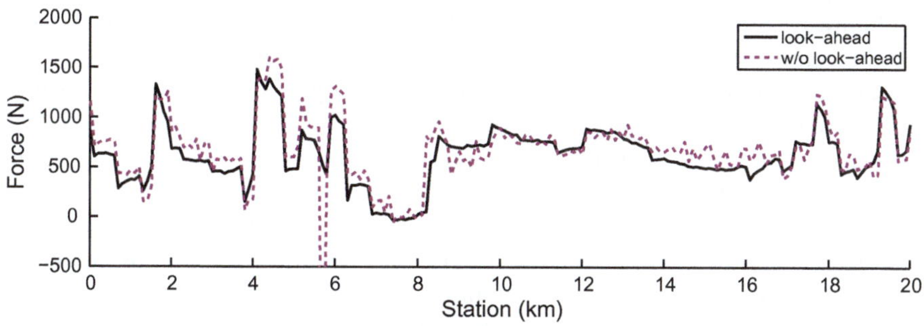

(b) Traction force

Fig. 6.8 Scenario 6: $q_{in} = 3000\ veh/h$, $\kappa = 50\%$, $R_{1,max} = 0.9$

Fig. 6.9 Typical form of $\mathrm{f}(q_{in}, \kappa)$ function

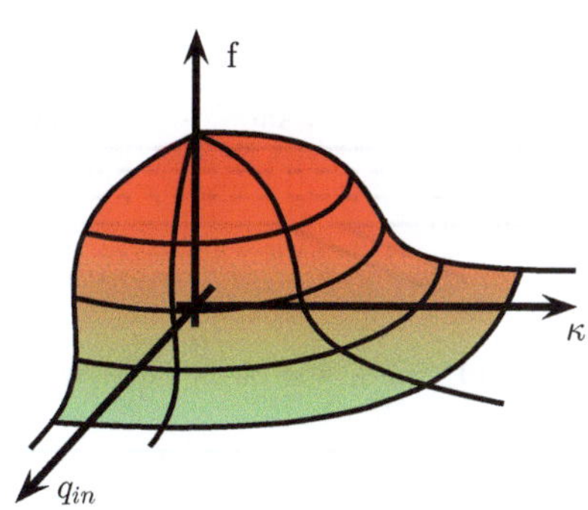

must be chosen as a function with low values. A typical form of the function $\mathrm{f}(q_{in}, \kappa)$, derived from the results of the analysis, is illustrated in Fig. 6.9. The function indicates that at low κ and N values the high $R_{1,max}$ value is preferred, which results in saving energy. If κ or q_{in} values increase, f is reduced. However, the inequality $R_{1,max} > 0$, $\forall \kappa, N$ must be guaranteed to improve the energy efficiency of the traffic flow.

6.4 Illustration of the Method

In this section, the enhancement of the traffic flow based on the improved predictive control strategy is illustrated. A traffic simulation scenario is presented in which the traffic volume along the highway section varies. The aim of the demonstration is to show that the predictive control vehicles are able to adapt to the variation of q. Thus, the energy efficiency of the entire traffic flow is improved.

The simulation scenario contains the analyzed 20 km section of the Hungarian M1 highway, see Fig. 6.10. At the 2 km point of this highway section, an entrance ramp is found, where vehicles merge into the highway. Furthermore, at 4 km an exit ramp is located. It results in an increased traffic on the section $2 \ldots 4$ km of the highway. Since q increases, $R_{1,max}$ of the predictive control must be reduced, see the control rule in Fig. 6.9. Thus the relationship between q and $R_{1,max}$ by a second-order polynomial is defined, see Fig. 6.11. It guarantees that the maximum of R_1 is reduced at increasing traffic volume.

In this section, three scenarios are presented. In the first case $R_{1,max} = 0.7$, which means that the faster motion of the vehicles is preferred. The second scenario $R_{1,max} = 0.9$ prefers the energy-efficient cruising, while in the third case $R_{1,max}$ is varied. The ratio of the predictive control vehicles in the traffic flow is constant $\kappa = 20\%$. Figure 6.12a presents the volume of the traffic for all simulations. It demonstrates the effect of the entrance and exit ramps, see the values of q in the $2 \ldots 4$ km section. As a result of the ramps, the traffic volume increases significantly in a short section, which may have an impact on the average speed of the traffic. Therefore, in the third simulation scenario the value of $R_{1,max}$ is varied based on the relationship

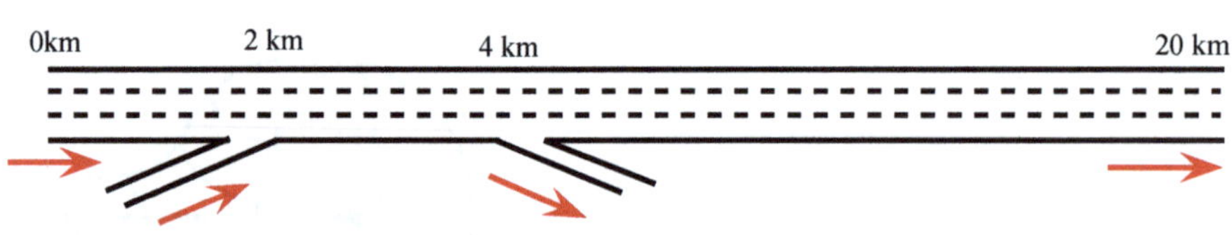

Fig. 6.10 Traffic flow on the highway section

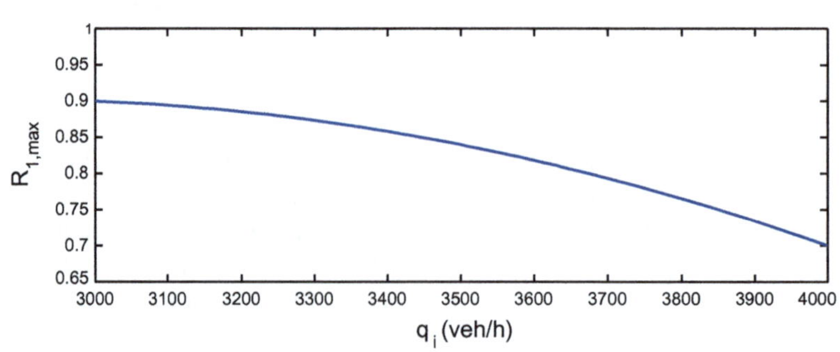

Fig. 6.11 Relationship between q and $R_{1,max}$

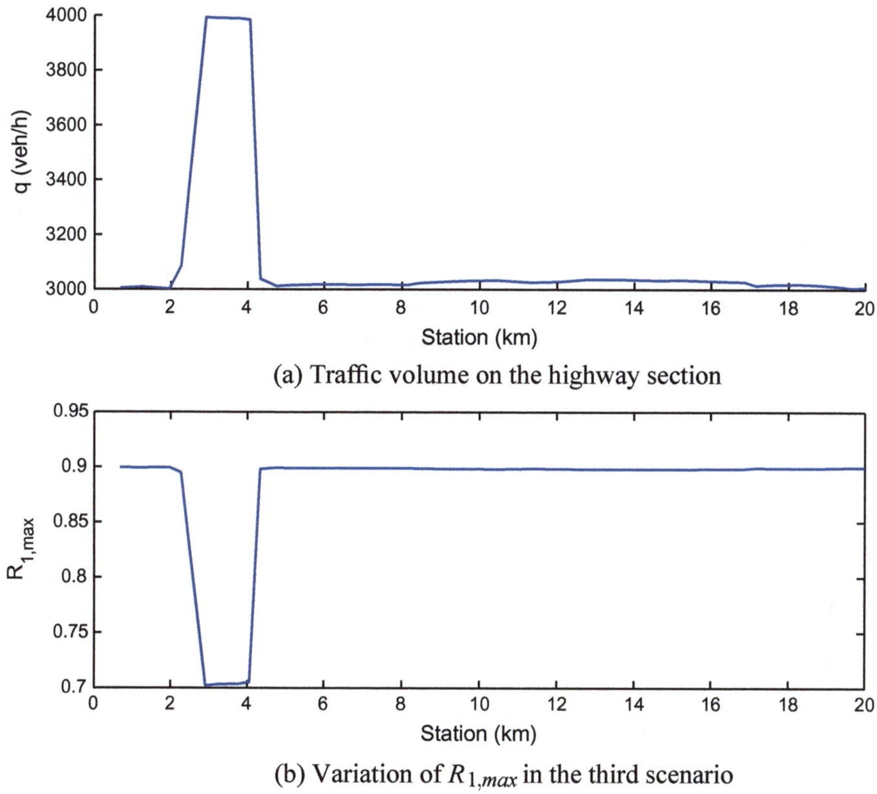

Fig. 6.12 Volume and $R_{1,max}$ values through the simulations

in Fig. 6.11. As a result, $R_{1,max}$ is reduced to 0.7 between 2 . . . 4 km. It means that in this 2 km long section, the priority of energy-efficient cruising is modified to faster motion.

Figure 6.13 presents the average speeds in the lanes, in which the predictive control vehicles have different $R_{1,max}$ strategies. If $R_{1,max} = 0.7$ is set, predictive control vehicles have a limited effect on the traffic. In this case, the average speed values are close to the maximum speed limit in all lanes. However, at the second strategy, $R_{1,max}$ has a high impact on the average speed, e.g. in the section 14 . . . 20 km. In this case, the speed of the vehicle varies significantly based on the forthcoming road terrain characteristics, speed limits, and the current traffic volume. Between the entrance and the exit ramps, the speed of the vehicles is unnecessary decreased, which is disadvantageous from the aspect of the traffic flow. The varying $R_{1,max}$ solution in the third scenario combines the advantages of the two solutions: it guarantees a fast speed on rush hour traffic section, while the energy-saving speed profile is realized on the further sections, see all of the lanes in Fig. 6.13. In this scenario, the $R_{1,max}$ is varied as Fig. 6.12b shows. For example, in Fig. 6.13a, the speed profile of the

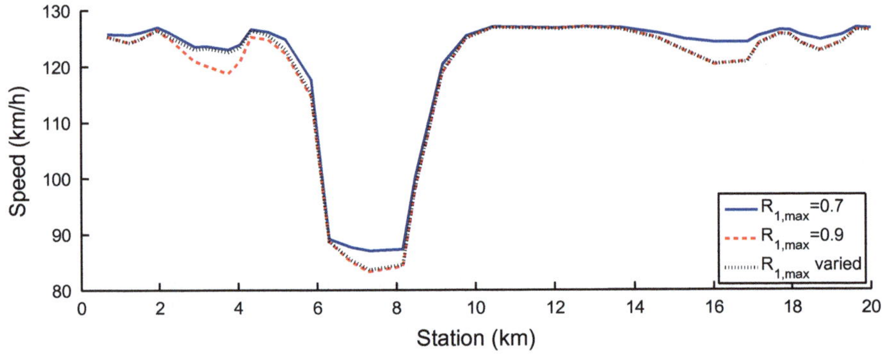

(a) Outer lane

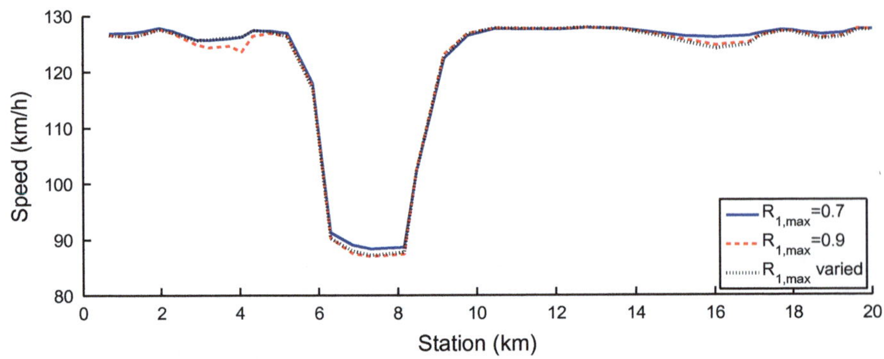

(b) Middle lane

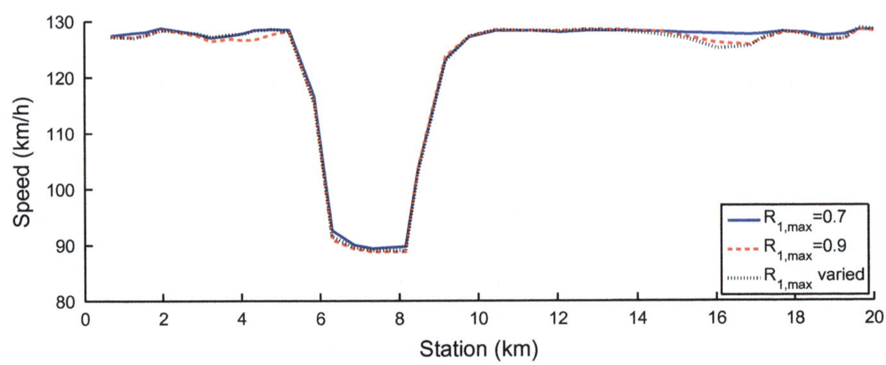

(c) Inner lane

Fig. 6.13 Average speed of the vehicles

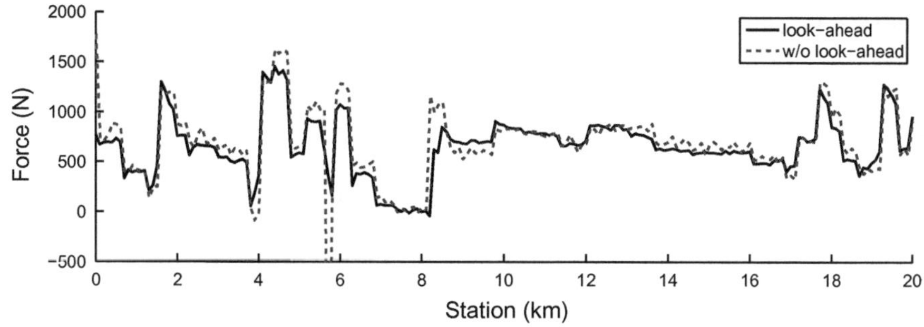

(a) $R_{1,max} = 0.7$

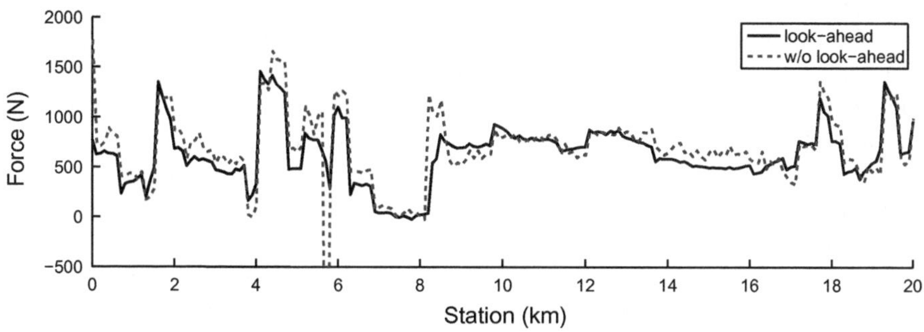

(b) $R_{1,max} = 0.9$

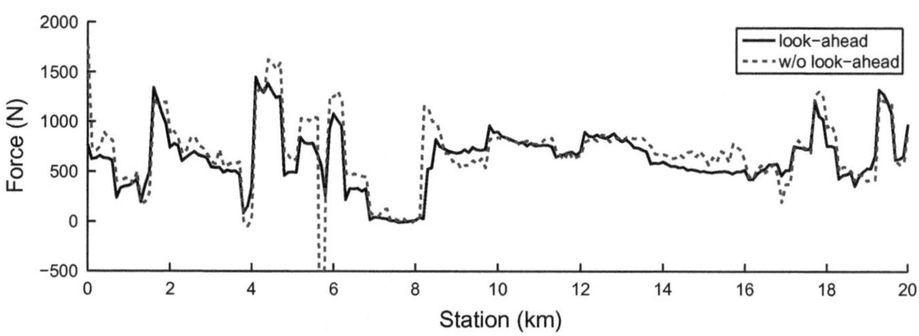

(c) $R_{1,max}$ varied

Fig. 6.14 Traction force of the vehicles

improved predictive control strategy is equal to the constant $R_{1,\,max} = 0.9$ scenario, except in the section $2 \ldots 4$ km, where the speed profile is close to the constant $R_{1,\,max} = 0.7$ strategy. Thus, it is possible to avoid the unnecessary deceleration in the rush hour traffic section, while the benefits of the predictive control are achieved.

The similar conclusions are drawn in the examination of the traction forces, see Fig. 6.14. In these figures, the dashed line is related to the traction forces of the

Table 6.2 Results of the simulations

κ(%)	$R_{1,max}$	Mean of force (N)			Traveling time (s)	Average speed (km/h)
		Conventional	Predictive	Overall		
20	0.7	675.5	629.0	666.2	611.8	117.8
20	0.9	673.3	598.7	658.4	612.0	117.6
20	Varied	676.5	604.7	662.1	611.9	117.7
1	0.7	677.7	621.8	677.1	609.1	118.2

conventional vehicles, while the solid line is connected to the predictive control vehicles. Figure 6.14c shows that the force values in the section 2 ... 4 km are close to the $R_{1,max} = 0.7$ strategy (Fig. 6.14a), while in the further sections, the forces are equal to those in the second strategy, see Fig. 6.14b. It reinforces the section between 14 km and 20 km points, where the similarity between the combined $R_{1,max}$ and the constant $R_{1,max} = 0.9$ strategy is demonstrated. Thus, the varying $R_{1,max}$ is able to guarantee reduced traction forces, while the average speed of the traffic is not degraded.

The numerical results of the simulations are found in Table 6.2. It proposes that in the $R_{1,max} = 0.9$ strategy the mean forces are smaller than in the first scenario $R_{1,max} = 0.7$. In the case of the varied $R_{1,max}$ scenario, the mean force of the predictive control vehicles is between the results of the constant scenarios, while the value for the conventional vehicles is slightly higher than the result of the $R_{1,max} = 0.7$. It is the consequence of the varying speed of the predictive control vehicles, which results in the accelerating and decelerating motion of the conventional vehicles in the section 2 ... 4 km, increasing the traction force. However, the strategy of the $R_{1,max}$ variation leads to the overall reduction of the mean forces compared to the $R_{1,max} = 0.7$ strategy. Moreover, comparing the results to the negligible number of the predictive control vehicles in the traffic flow (scenario $\kappa = 1\%$) the benefits are more significant, see Table 6.2. Through the use of the predictive control strategy the force saving of the overall traffic is 2.2%. Thus, the proposed predictive control strategy can be effective in the traffic flow, where the predictive control vehicles are in minority.

References

Bekiaris-Liberis N, Roncoli C, Papageorgiou M (2016) Highway traffic state estimation with mixed connected and conventional vehicles. IEEE Trans Intell Transp Syst 17(12):3484–3497

Bevly D, Cao X, Gordon M, Ozbilgin G, Kari D, Nelson B, Woodruff J, Barth M, Murray C, Kurt A, Redmill K, Ozguner U (2016) Lane change and merge maneuvers for connected and automated vehicles: A survey. IEEE Trans Intell Veh 1(1):105–120

Bose A, Ioannou P (2003) Analysis of traffic flow with mixed manual and semi-automated vehicles. IEEE Trans Intell Transp Syst 4(4):173–188

Chin H, Okuda H, Tazaki Y, Suzuki T (2015) Model predictive cooperative cruise control in mixed traffic. In: IECON 2015—41st annual conference of the IEEE industrial electronics society, pp 3199–3205

Gáspár P, Németh B (2015) Design of look-ahead cruise control using road and traffic conditions. In: American control conference, Chicago, USA, pp 3447–3452

Jerath K, Ray A, Brennan S, Gayah VV (2015) Dynamic prediction of vehicle cluster distribution in mixed traffic: A statistical mechanics-inspired method. IEEE Trans Intell Transp Syst 16(5):2424–2434

Jia D, Lu K, Wang J, Zhang X, Shen X (2016) A survey on platoon-based vehicular cyber-physical systems. IEEE Commun Surv Tutor 18(1):263–284

Johannesson L, Nilsson M, Murgovski N (2015) Look-ahead vehicle energy management with traffic predictions. In: IFAC Workshop on engine and powertrain control, simulation and modeling, columbus, Ohio, USA

Li K, Ioannou P (2004) Modeling of traffic flow of automated vehicles. IEEE Trans Intell Transp Systems 5(2):99–113

Ngoduy D, Liu R (2007) Multiclass first-order simulation model to explain nonlinear traffic phenomena. Phys A Statist Mech Appl 385(2):667–682

Schakel W, Van Arem B, Netten B (2010) Effects of cooperative adaptive cruise control on traffic flow stability. In: 2010 13th International IEEE conference on intelligent transportation systems (ITSC), pp 759–764, https://doi.org/10.1109/ITSC.2010.5625133

Talebpour A, Mahmassani HS (2015) Influence of autonomous and connected vehicles on stability of traffic flow. In: Transportation research board 94th annual meeting, Washington DC, United States

van Arem B, van Driel CJG, Visser R (2006) The impact of cooperative adaptive cruise control on traffic-flow characteristics. IEEE Trans Intell Transp Systems 7(4):429–436

Wageningen-Kessels F (2013) Multi-class continuum traffic flow models: analysis and simulation methods. PhD thesis, Delft University of Technology, Ph.D. thesis

Zhang C, Vahidi A (2011) Predictive cruise control with probabilistic constraints for eco driving. In: ASME 2011 dynamic systems and control conference and bath/ASME symposium on fluid power and motion control, vol 2, pp 233–238

Part III
Control Strategies

Chapter 7
Control Strategy of the Ramp Metering in the Mixed Traffic Flow

Introduction and Motivation

In, the last decade the conception of the autonomous control of the road vehicles has been in a focus of automotive research activities. Autonomous vehicles may have an impact on the traffic flow, which results from the smooth driving and adhering to the traffic regulations. Moreover, the automation of the vehicles requires intelligent transportation systems with information about the traffic environment.

The autonomous vehicle control facilitates the communication between the vehicles and the infrastructure. The coordination of the vehicles is able to reduce energy consumption and improve the efficiency of their motion in the traffic. A multiple-stack architecture of vehicle cooperation with different layers is proposed in pon Lin and Maxemchuk (2012); Wei and Dolan (2009); Halle and Chaib-draa (2005). Simulation results show the positive impacts of the layer-based communication and control on the stability of the traffic flow, see e.g., Monteil et al. (2013). A hierarchical decomposition strategy for handling context information is proposed in Fuchs et al. (2007).

In this chapter, the macroscopic traffic flow modeling and the ramp metering control of the traffic dynamics and the optimization of the look-ahead cruise control of the microscopic vehicles are coordinated. As a novelty of the approach, the impact of the look-ahead cruise control on the traffic flow is analyzed. The control of the individual vehicles and the traffic control are handled simultaneously, consequently, a trade-off between the parameters of the microscopic and the macroscopic models has been created. The contribution will be an optimization strategy, which incorporates the nonlinearities and the parameter-dependency of the traffic system. The coordination of the traffic and the vehicles is presented in a highway, in which there are two intervention possibilities such as the control of the input traffic flow on the highway gates and the parameter setting of the look-ahead cruise control method. The appropriate selection of these control signals guarantees the avoidance of the

P. Gáspár and B. Németh, *Predictive Cruise Control for Road Vehicles Using Road and Traffic Information*, Advances in Industrial Control,
https://doi.org/10.1007/978-3-030-04116-8_7

congestion, the minimization of the vehicle energy consumption and the reduction of the queue length at the controlled gates.

7.1 Modeling the Effect of Cruise Controlled Vehicles on Traffic Flow

The basic and most important law in a traffic network is the law of conservation, which is formulated as follows:

$$\rho_i(k+1) = \rho_i(k) + \frac{T}{L_i}\left[q_{i-1}(k) - q_i(k) + r_i(k) - s_i(k)\right] \tag{7.1}$$

where k is the discrete time step index, T is the discrete sample time step, ρ_i, q_i are the traffic density [veh/km], and flow [veh/h] of the segment i, respectively, r_i and s_i are the sum of ramp inflow and outflow values in the units [veh/h], q_{i-1} is the flow in segment $i-1$, which is also an inflow of segment i. The length of the segment is denoted by L_i.

In 9.1, the outflows $q_{i-1}(k)$ and $s_i(k)$ are measured values, while $r_i(k)$ is the controlled inflow. However, the outflow $q_i(k)$ of segment i depends on several factors, see e.g., Messmer and Papageorgiou (1990); Treiber and Kesting (2013). The determination of the outflow is based on the fundamental relationship $q_i(k) = \rho_i(k)v_i(k)$, where $v_i(k)$ is the flow speed Ashton (1966). In practice, the relationship between $q_i(k)$ and $\rho_i(k)$ is characterized by using historic measurements, see Gartner and Wagner (2008), as follows:

$$q_i(k) = \mathscr{F}(\rho_i(k)) \tag{7.2}$$

It is important to note that the exact knowledge of the complete operational fundamental diagram assumes the measurements of all the above traffic variables, see Papageorgiou and Vigos (2008). Thus, the traffic model is transformed as

$$\rho_i(k+1) = \rho_i(k) + \frac{T}{L_i}\left[\mathscr{F}(\rho_i(k)) - q_i(k) + r_i(k) - s_i(k)\right] \tag{7.3}$$

The traffic system model and the measurements of a test network are built in VisSim. As a demonstration example, a 3-lane 20km-long segment of the Hungarian M1 highway between Budapest and Tatabánya is modeled in VisSim, in which the terrain characteristics (see Fig. 7.1), the speed limits and the three number of lanes are also built. The speed limit in the section is 130 km/h, and there is a 90 km/h further limit in the section 5.6 km ... 8.5 km. Using this model, several simulations with different traffic density are performed. Moreover, the traffic flow contains a significant number of look-ahead vehicles, whose speed selections are fitted to the forthcoming slopes and speed limits.

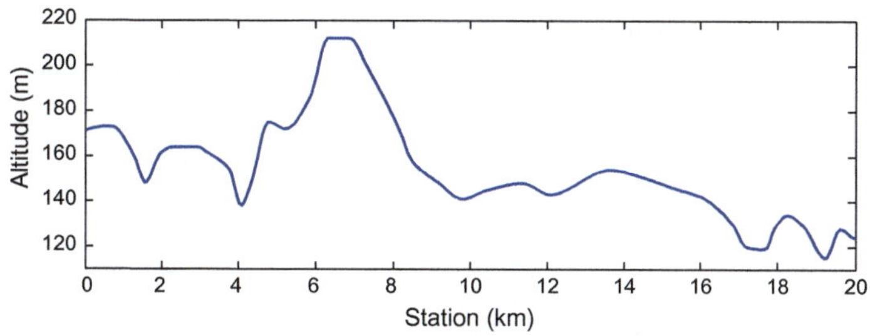

Fig. 7.1 Terrain characteristics of the highway section

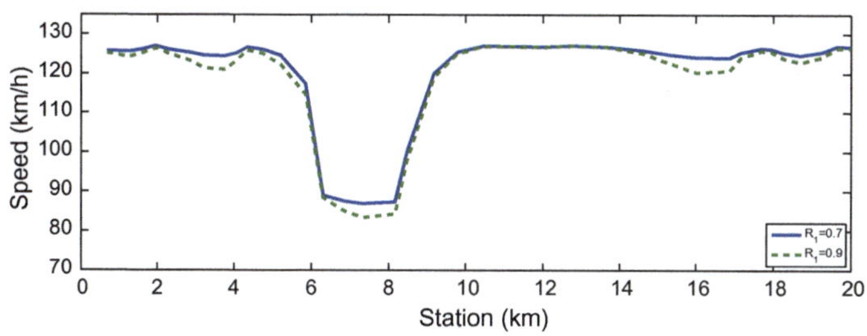

Fig. 7.2 Flow speed on the road

The computation of the current speed is based on a multi-objective optimization, in which the minimization of the required longitudinal energy (or fuel consumption) and the reduction of the traveling time are involved, see Németh and Gáspár (2013b). The two optimization criteria lead to different optimum solutions. In order to create a balance between the performance requirements, an optimization parameter R_1 is introduced. This parameter significantly determines the speed characteristics: If R_1 is set its minimum, the minimization of the traveling time is enforced. However, if R_1 is set its maximum, it yields energy-efficient cruising. The appropriate selection of R_1 has an impact on the traffic flow.

The speed profile changes during traveling according to the forthcoming terrain characteristics and speed limits. In the traffic scenario, two types of vehicles are considered:

- vehicles without look-ahead control, which track the current maximum speed limit and ignore energy demands,
- vehicles with look-ahead control, which track their own energy/time optimal speed profile.

In the simulations of the traffic flow, the rate of the look-ahead vehicles is 20%. Consequently, the cruising of the look-ahead vehicles has a significant effect on the overall traffic flow. Figure 7.2 demonstrates the speed of the traffic flow in the highway section.

Moreover, R_1 also has an effect on the energy consumption of the entire traffic flow. Table 7.1 shows the results in the mean of the longitudinal forces of the conventional

Table 7.1 Results in the longitudinal forces

R_1	$F_{conv}[N]$	$F_{look}[N]$	Energy consumption (%)
0.7	676.2	629.4	7.2
0.9	674.8	598.7	11.9

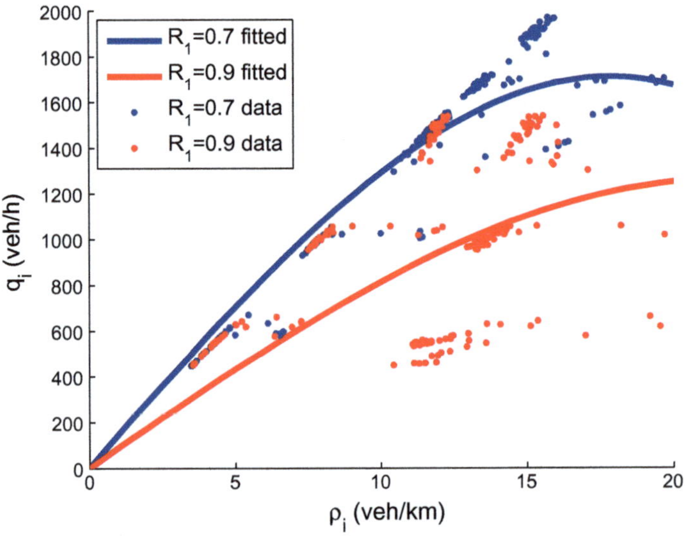

Fig. 7.3 Fundamental diagram of the traffic network

vehicles F_{conv} and the look-ahead vehicles F_{look} with their energy consumption. The table shows that the control forces of the look-ahead vehicles are significantly reduced when the optimization parameter R_1 is increased. However, R_1 has a slight impact on the conventional vehicles. Furthermore, the energy consumption of all the vehicles decreases significantly, which is the result of the interconnection between the vehicles in the traffic flow.

Figure 7.3 illustrates an example about the measurements and the fundamental diagram $q_i(k) = \mathscr{F}(\rho_i(k))$ of the system. The simulations are performed with different traffic densities and different optimization parameters in the look-ahead control, $R_1 = 0.7$ and $R_1 = 0.9$. Polynomial functions are fitted to the different scenarios, as shown in Fig. 7.3. The results illustrate that R_1 has a significant impact on the fundamental diagram. If R_1 is increased, then the outflow q_i is reduced due to the decreased vehicle speed. However, the critical density $q_{i,crit}$, which is related to the maximum of the fundamental diagram, is higher. Thus, the selection of R_1 has an impact on both the outflow and the critical density, thus also on the stability of the traffic flow.

The analysis is suitable for the determination of the function $\mathscr{F}$ in fundamental relationship (7.2). A parameter-dependent polynomial function is formulated, which considers the effect of R_1 variation. The fundamental relationship between $\rho_i(k)$ and $q_i(k)$ is formed as

$$q_i(k) = \sum_{j=1}^{n} c_j(R_1(k))\rho_i(k)^j \tag{7.4}$$

where the coefficients in the polynomial function are formed as

$$c_j(R_1(k)) = \sum_{l=1}^{m} d_l R_1(k)^l, \tag{7.5}$$

in which d_l are constants and m is the number of the members.

In a summary, the applied traffic network model is formulated based on the law of conservation (7.3) and the fundamental relationship (7.4). The state-space equation of the system is given as follows:

$$x(k+1) = f(x(k), R_1(k)) + gd_i(k) + gr_i(k) \tag{7.6}$$

where $x(k) = \rho_i(k)$ is the state of the system, $f(x(k), R_1(k))$ is a polynomial parameter-dependent function, which describes the dynamics of the traffic. The measured disturbance are as follows:

$$d_i(k) = \left[q_{i-1}(k)\ s_i(k)\right]^T. \tag{7.7}$$

The control input is the inflow on the controlled gates $r_i(k)$. Note that in the state-space representation the optimization parameter $R_1(k)$ is also a control value. Thus, the dynamics of the traffic can be influenced through the signals $r_i(k)$ and $R_1(k)$.

7.2 Stability Analysis of the Traffic System

In, this section the stability of the traffic system based on the SOS method is analyzed. The purpose of the analysis is to examine how the variation of the optimization parameter $R_1(k)$ and the control inflow $u(k) = r_i(k)$ influence the stability of the system. It is also necessary to determine what the maximum set $u_{max}(R_1(k))$ is, by which the congestion of the network can be avoided. The function $u_{max}(R_1(k))$ also results in the relationship between the control interventions. Since the measured disturbance $d_i(k)$ has an effect on the highway network dynamics, these values must be handled together with the set $u_{max}(R_1(k))$. In the following, the application of the SOS method on the controlled invariant set computation is shown, while the theoretical background and the details can be found in Appendix B.

The state-space representation of the system (7.6) with $u_{max}(R_1(k))$ is given in the following form:

$$x(k+1) = f(R_1(k), x(k)) + gu_{max}(R_1(k)) \tag{7.8}$$

where $f(R_1(k), x(k))$ is a matrix, which incorporates smooth polynomial functions and $f(R_1, 0) = 0$.

The parameter-dependent Control Lyapunov Function is chosen in the following form:

$$\mathbf{V}(R_1(k), x(k)) = V(x(k)) \cdot b(R_1(k)) \tag{7.9}$$

where $b(R_1(k))$ is an intuitively chosen R_1-dependent basis function. The difference in the condition (B.29) is expanded using (B.30)

$$\begin{aligned}\Delta\mathbf{V}(R_1(k), x(k)) =& \mathbf{V}(x(k+1), R_1(k)) - \mathbf{V}(x(k), R_1(k)) + \\ &+ \mathbf{V}(x(k), R_1(k+1)) - \mathbf{V}(x(k), R_1(k)) \\ =& (V(x(k+1)) - V(x(k))) \cdot b(R_1(k)) + \\ &+ (b(R_1(k+1)) - b(R_1(k))) \cdot V(x(k)) \\ =& \Delta V(x(k)) \cdot b(R_1(k)) + \Delta b(R_1(k)) \cdot V(x(k))\end{aligned} \tag{7.10}$$

During the decomposition of $x(k)$ and $R_1(k)$ in the Control Lyapunov Function, the original difference $\Delta\mathbf{V}(R_1(k), x(k))$ is separated to $\Delta V(x(k))$ and $\Delta b(R_1(k))$. The difference $\Delta V(x(k))$ using (B.28) is computed as

$$\Delta V(x(k)) = V(f(R_1(k), x(k)) + gu(k)) - V(x(k)) \tag{7.11}$$

Considering the function $u_{max}(R_1(k))$ in (B.32)

$$\Delta V(x(k))\Big|_{u(k)=u_{max}(R_1(k))} = V(f(R_1(k), x(k)) + gu_{max}(R_1(k))) - V(x(k)) \tag{7.12}$$

The difference $\Delta b(R_1(k))$ depends on the intuitively formulated basis function $b(R_1(k))$. In the further computation, the upper limit of $\Delta b(R_1(k))$ is used such as $\nu \geq |\Delta b(R_1(k))|$.

The local stability criterion determines the set of the states x, which are stable, and yields that $x_\infty < \infty$, if u_{max} is applied. This set is bounded by the Controlled Invariant Set and is defined in the following way: $\mathbf{V}(R_1(k), x(k)) = 1$. For practical reasons the equality definition is substituted for by two inequality conditions:

$$1 - \varepsilon \leq \mathbf{V}(R_1(k), x(k)) \leq 1 \tag{7.13}$$

where $\varepsilon > 0$ is an infinitesimally small number.

Further constraints on the stabilization are the validity ranges of the scheduling variable R_1 and the state x. The choice of the basis function $b(R_1(k))$ is valid in a range

$$R_{1,min} \leq R_1(k) \leq R_{1,max} \tag{7.14}$$

Moreover, the solution of the urban network gating must be found at the constraint:

$$0 \leq x(k) \tag{7.15}$$

which represents that the traffic density is positive or zero.

The optimization problem is to find an $u_{max}(R_1(k))$ solution and feasible $\mathbf{V}(R_1(k), x(k))$ for the following task:

$$\max \; u_{max}(R_1(k)) \tag{7.16}$$

over $s_1, s_2, s_3, s_4, s_5 \in \Sigma_n;\; V(x(k)), b(R_1(k)) \in \mathscr{R}_n$
such that

$$\begin{aligned} &-\Big((V(f(R_1(k), x(k)) + gu_{max}(R_1(k)))- \\ &- V(x(k))) \cdot b(R_1(k)) + \nu \cdot V(x(k))\Big)- \\ &- s_1\Big(V(x(k)) \cdot b(R_1(k)) - (1-\varepsilon)\Big)- \\ &- s_2\Big(1 - V(x(k)) \cdot b(R_1(k))\Big) - s_3 x(k)- \\ &- s_4\left(R_1(k) - R_{1,min}\right) - s_5\left(R_{1,max} - R_1(k)\right) \in \Sigma_n \end{aligned} \tag{7.17}$$

The result of the optimization defines the maximum Controlled Invariant Set, in which the system is stable under the function $u_{max}(R_1(k))$. Since $u(k) = r_i(k)$ and the parameter $R_1(k)$ are the control interventions of the system, $u_{max}(R_1(k))$ results in the relationship between them. The computation method of the maximum Controlled Invariant Set is found in Appendix B.

7.3 Control Strategy of the Ramp Metering and the Cruise Controlled Vehicles

In, this section the selection of the control interventions, such as ramp metering command $r_i(k)$ and the look-ahead parameter $R_1(k)$, is derived. The function $u_{max}(R_1(k))$ determines the region of the stability and there are several intervention possibilities. Two performance requirements, which must be guaranteed by the analysis-based control strategy, are defined as follows:

- The energy consumption of the vehicles during the journey must be minimized. There is a relationship between the look-ahead parameter $R_1(k)$ and the energy consumption. The form of the first performance is

$$R_1(k) \rightarrow \text{Max}. \tag{7.18}$$

- The length of the queue on the inflow ramps must be reduced, which results in the reduction of the waiting time of the inflow vehicles. The length of the queues at the controlled gates is the following:

$$l_i(k+1) = l_i(k) + T\left(r_{i,dem}(k) - r_i(k)\right) \tag{7.19}$$

  where $l_i(k)$ denotes the queue length in the unit [veh], $r_i(k)$ is the ramp inflow and $r_{i,dem}(k)$ is the demand of the ramp inflow. The control must guarantee the following criterion: $l_i(k) \leq l_{max}(k)$, where $l_{max}(k)$ is the maximum acceptable length in the queue.

The two performance requirements lead to an optimization task. Its solution yields the optimum $r_i(k)$ and $R_1(k)$. Since in the computation of $u_{max}(R_1(k))$ all of the inflow vehicles are considered, it is also necessary to consider the measured disturbance vehicles $d_i(k)$. The optimization task is as follows:

$$\max \ R_1(k) \tag{7.20}$$

over $r_i(k)$; $R_1(k)$ such that

$$r_i(k) + d_i(k) \leq u_{max}(R_1(k)) \tag{7.21}$$

$$l_i(k) \leq l_{max}(k) \tag{7.22}$$

$$R_{1,min} \leq R_1(k) \leq R_{1,max} \tag{7.23}$$

$$0 \leq r_i(k) \tag{7.24}$$

The result of the optimization (7.20) leads to the actual $r_i(k)$ and $R_1(k)$ values. Generally, the form of the function $u_{max}(R_1)$ is linear in $R_1(k)$ and the computation of (7.20) can be calculated in an efficient way. In the implementation of the method, only the optimization (7.20) must be calculated in an online way. The complex optimization $u_{max}(R_1(k))$ (B.39) is solved in advance, and only its result is used in the optimization $r_i(k)$, $R_1(k)$.

7.4 Simulation Results

In this section, the SOS-based analysis and control design are illustrated through a simulation example. The goals of the simulation are to show that the set-based analysis yields a maximum stabilizable control input, and to design the optimal

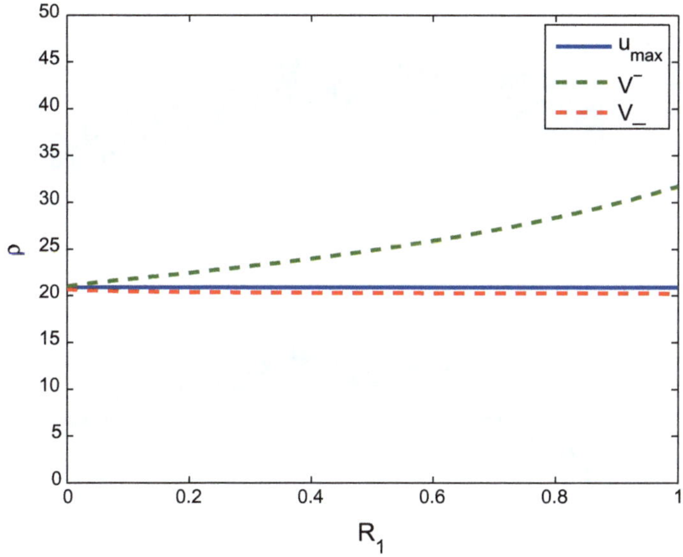

Fig. 7.4 Maximum controlled invariant set of the system

control solution. In the simulation example, a 20km-long segment of the Hungarian M1 highway between Budapest and Tatabánya is considered, see Sect. 7.1. Both the nonlinear stability analysis and the simulation are based on the VisSim software.

First, the results of the stability analysis are presented. The Maximum Controlled Invariant Set, which is computed through the SOS method and the criterion (B.39), is found in Fig. 7.4. It shows the computed optimum function using $u_{max} = 2073 - 592R_1$ with its boundaries $\mathbf{V}^-$ and $\mathbf{V}_-$. If the traffic density is between the boundary states determined by $\mathbf{V}^-$ and $\mathbf{V}_-$, the system is stable. Moreover, the form u_{max} represents that the increase in R_1 reduces the flow of the vehicles which can be injected into the traffic. For example, if $R_1 = 0.5$, then $u_{max} = 1777$ veh/h, while at $R_1 = 0.9$ the maximum input vehicle flow u_{max} is around 1540 veh/h. It yields 13% reduction in the input flow.

Second, the application of the control variables $r_i(k)$ and $R_1(k)$ is presented through a simulation scenario. Figure 7.5a shows the d_i disturbance flow of the highway section. Moreover, there is also an input flow demand on the controlled gates $r_{i,dem}$, as shown in Fig. 7.5b. In the first part of the simulation, the flow of the vehicles is increased and in the second part, it is reduced. Since the number of the vehicles varies randomly during the time, the flows are modeled as stochastic signals.

The control interventions are shown in Fig. 7.5c and d. In the first 80 min $r_i(k) = r_{i,dem}(k)$, because $r_i(k) + d_i(k) < u_{max}(R_1(k))$. Moreover, the optimization parameter is set $R_1(k) = 1$, see Fig. 7.5d. Between the time 80 ... 100 min, the actuation $r_i(k)$ is limited by u_{max}, see Fig. 7.5c. It means that the disturbance and $r_{i,dem}(k)$ are sufficiently large, and the length of the queue increases, see Fig. 7.6b. However, the increase in the queue length can be considered as a puffer, which results that

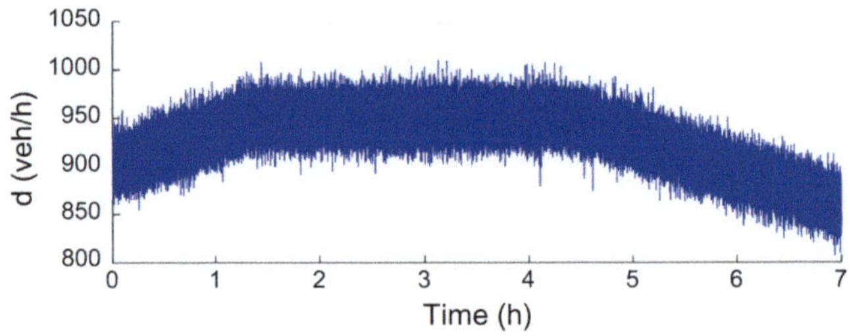

(a) Disturbance flow, $d_i(k)$

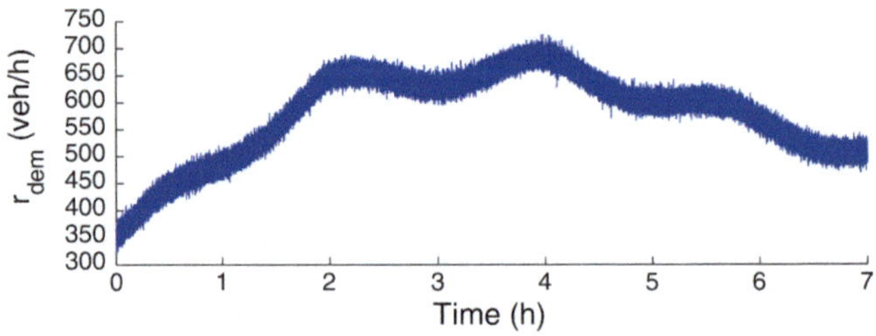

(b) Input flow demand on the controlled gates, $r_{i,dem}(k)$

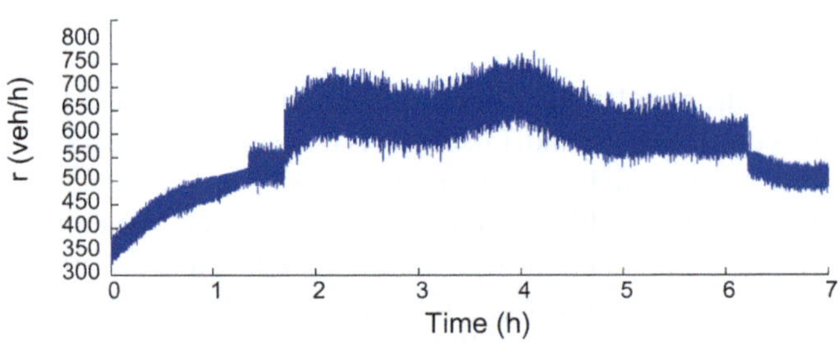

(c) Control flow, $r_i(k)$

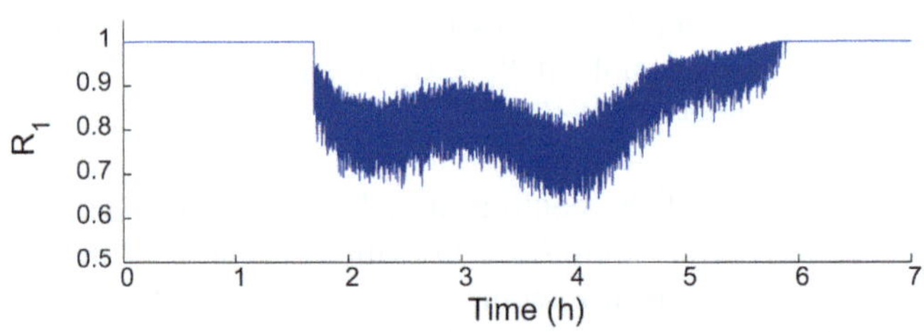

(d) Optimum of the look-ahead parameter, $R_1(k)$

Fig. 7.5 Simulation scenario

the reduction of $R_1(k)$ is not necessary. At 100 min the length of the queue reaches the limit $l_{max}(k) = 10$, which yields that $R_1(k)$ must be reduced based on the optimization process (7.20). The decrease in $R_1(k)$ leads to the increase in the maximum

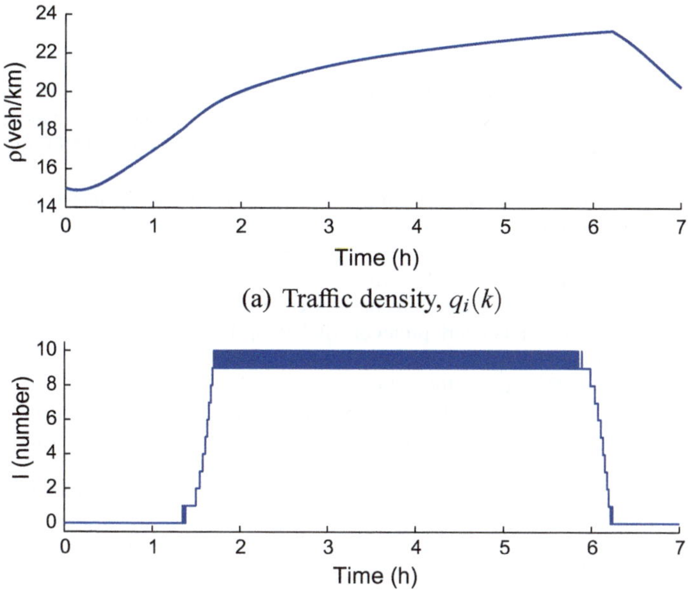

(a) Traffic density, $q_i(k)$

(b) Queue length on the controlled gates, $l_i(k)$

Fig. 7.6 Simulation results

traffic flow, as illustrated in the fundamental diagram Fig. 7.3. Around 6 h the inflows $d_i(k)$ and $r_{i,dem}(k)$ decrease, which results in the reduction of the queue length, see Fig. 7.6b.

The traffic density shows that the stability of the system is guaranteed throughout the entire simulation, see Fig. 7.6a. The proposed control algorithm is suitable for the coordination of the traffic flow and the optimization setting of the look-ahead vehicles.

In the chapter, the design strategy of the traffic flow with vehicles using look-ahead cruise control methods has been proposed. A model of the traffic system has been extended with the optimization parameter of the look-ahead vehicles. A two-step optimization method for the control design has been proposed. The stability of the system has been examined by using the maximum Controlled Invariant Set of the system $u_{max}(R_1(k))$ and the optimization criterion (B.39). The computation of the control interventions $r_i(k)$, $R_1(k)$ has been performed by using a multicriteria optimization task (7.20). The balance between the reduction of the energy demands of the traffic flow and simultaneously the reduction of the queue length of the controlled lanes can be achieved. In the implementation of the method, only the optimization of the control intervention must be calculated in an online way. The complex optimization $u_{max}(R_1(k))$ is solved in advance, and only its result is used in the optimization $r_i(k)$, $R_1(k)$. The proposed method has several advantages such as the simple implementation, handling the system nonlinearities, considering the parameter dependences.

References

Ashton W (1966) The theory of traffic flow. Spottiswoode, Ballantyne and Co. Ltd., London

Fuchs S, Rass S, Lamprecht B, Kyamakya K (2007) Context-awareness and collaborative driving for intelligent vehicles and smart roads. In: 1st International workshop on ITS for an ubiquitous ROADS, Los Alamitos

Gartner N, Wagner P (2008) Analysis of traffic flow characteristics on signalized arterials. Transp Res Record 1883:94–100

Halle S, Chaib-draa B (2005) A collaborative driving system based on multiagent modelling and simulations. Transp Res Part C 13:320–345

Lin SP, Maxemchuk NF (2012) An architecture for collaborative driving systems. In: 20th IEEE International conference on network protocols (ICNP) pp 1–2

Messmer A, Papageorgiou M (1990) METANET—a macroscopic simulation program for motorway networks. Traffic Eng Control 31:466–470

Monteil J, Billot R, Sau J, Armetta F, Hassas S, Faouzi NEE (2013) Cooperative highway traffic: multiagent modeling and robustness assessment of local perturbations. J Transp Res Board 2391:1–10

Németh B, Gáspár P (2013b) Design of vehicle cruise control using road inclinations. Int J Veh Auton Syst 11(4):313–333

Papageorgiou M, Vigos G (2008) Relating time-occupancy measurements to space-occupancy and link vehicle-count. Transp Res Part C 16(1):1–17

Treiber M, Kesting A (2013) Traffic flow dynamics: data models and simulation. Springer, Berlin

Wei J, Dolan JM (2009) A multi-level collaborative driving framework for autonomous vehicles. In: IEEE international symposium robot and human interactive communication, Toyama, Japan pp 40–45

Chapter 8
MPC-Based Coordinated Control Design of the Ramp Metering

Introduction and Motivation

The modeling and analysis of the mixed traffic, in which controlled vehicles and conventional uncontrolled vehicles are traveling together, require novel methodologies. Since the speed profiles of the look-ahead vehicles may differ from those of the conventional vehicles, the characteristics of the traffic flow change. As motivations, two examples from the previous research results of the authors are presented. In practice, the relationship between the macroscopic traffic flow and the traffic density forms the basis of the fundamental diagram. Figure 7.3 shows the diagram changes with the selection of the energy-efficient scaling parameter R_1. Since in the optimization problem there is a trade-off between the minimization of the force/energy and that of the traveling time, an energy-efficient parameter R_1 is introduced. If R_1 selected is large, the control focuses on the minimization of the longitudinal force but if R_1 is small, the control focuses on traveling time. If this parameter is increased, then the traffic flow decreases due to the decreased vehicle speed. However, the critical density $q_{i,crit}$, which is related to the maximum traffic flow, is higher. The selection of R_1 has an impact on both the outflow and the critical density, thus also on the stability of the traffic flow. A detailed analysis is found in Németh et al. (2017a).

The following analysis presents the impact of the parameter R_1 on the traffic flow. A 20-km-long, 3-lane segments of the Hungarian M1 highway is modeled in VisSim as functions of the traffic inflow of the vehicles (q_{in}) and the ratio of the look-ahead vehicles in the entire traffic (κ). The following conclusions on the traction forces based on the simulations are drawn. More details are found in Németh et al. (2017b).

- The increase in q_{in} can lead to the reduction of the traction forces of the conventional and the look-ahead vehicles, simultaneously. However, it increases the traveling time of the vehicles due to the dense traffic.
- The increase in κ slightly increases the required force of the look-ahead vehicles, but it reduces it significantly in conventional vehicles.

P. Gáspár and B. Németh, *Predictive Cruise Control for Road Vehicles Using Road and Traffic Information*, Advances in Industrial Control,
https://doi.org/10.1007/978-3-030-04116-8_8

- The increase in $R_{1,max}$ leads to the significant reduction of the forces of the look-ahead vehicles and it has high impact on the conventional vehicles.

A method will be developed in which the control of the macroscopic traffic flow and the cruise control of the local vehicles are coordinated. The contribution will be an optimization strategy, which incorporates the nonlinearities and the parameter dependency of the traffic system and the multi-optimization of the look-ahead vehicles. Consequently, a trade-off between the parameters of the microscopic and the macroscopic models has been created. In the modeling part, the Sum-of-Squares (SOS) programming method, while in the control design the Model Predictive Control (MPC) are applied. The role of the MPC control is to consider the prediction of the traffic flow in the design. A robust MPC method was proposed to urban traffic systems in Tettamanti et al. (2014), Frejo and Camacho (2012).

The coordination of the traffic system and the vehicles is presented on a highway. In the mixed traffic system, vehicles equipped with look-ahead control and conventional uncontrolled vehicles are traveling together. In this task, there are two intervention possibilities such as the ramp metering control on the highway gates and the parameter setting of the look-ahead cruise control. In a previous paper, a control design method to guarantee stability for only one highway section with a fixed energy-efficient parameter was proposed, see Németh et al. (2017a). The purposes of the coordinated control are to improve the energy efficiency of the entire traffic flow and simultaneously stabilize the controlled highway, reduce the queue length at the controlled gates, and avoid congestion. The modeling and the stability analysis are based on the SOS programming method. In the mixed traffic system, vehicles equipped with look-ahead control and conventional uncontrolled vehicles are traveling together. The result of the analysis is the maximum controlled invariant set of the traffic flow and an inequality with the inflow and outflow of the system. In the control design, the MPC method is applied, by which the prediction of the traffic flow and that of the traveling of the vehicles are taken into consideration. The traffic system is considered to have ramp metering control, which has a significant effect on the inflow on the highway. In the coordinated control, the energy-efficient parameter is built into the control design in order to guarantee the dynamics and stability of the traffic system besides reducing the force/energy of the individual vehicles.

8.1 Modeling and Analysis of the Traffic Flow with Cruise Controlled Vehicles

The modeling of traffic dynamics in highway systems is based on the law of conservation. The relationship contains the sum of inflows and the outflows for a given highway segment i. The traffic density ρ_i $[veh/km]$ is expressed in the following way:

$$\rho_i(k+1) = \rho_i(k) + \frac{T}{L_i}\left[q_{i-1}(k) - q_i(k) + r_i(k) - s_i(k)\right], \tag{8.1}$$

where k denotes the index of the discrete time step, T is the discrete sample time, L is the length of the segment, q_i $[veh/h]$ and q_{i-1} $[veh/h]$ denote the inflow of the traffic in segments i and $i-1$, and r_i $[veh/h]$ and s_i $[veh/h]$ are the sum of ramp inflow and outflow values, respectively. In Eq. (8.1), the inflow $q_{i-1}(k)$ and the ramp outflow $s_i(k)$ are measured disturbance values, while $r_i(k)$ is a controlled ramp metering inflow, which is introduced by Papageorgiou and Vigos (2008). The outflow $q_i(k)$ of segment i incorporates the core of the traffic dynamics and depends on several factors, see e.g. Messmer and Papageorgiou (1990), Treiber and Kesting (2013). In the following, $q_i(k)$ is formulated according to the fundamental relationship, see Ashton (1966), which is generally derived through historic measurements, see Gartner and Wagner (2008), such as

$$q_i(k) = \mathscr{F}(\rho_i(k)). \tag{8.2}$$

Note that $q_i(k)$ is not only the outflow of segment i, but it is also the inflow of segment $i+1$. Although $q_i(k)$ is measured as the inflow of segment $i+1$, in the modeling it is required to consider it through the formula Eq. (8.2) to receive signals from the dynamics of the traffic flow.

The function $\mathscr{F}(\rho_i(k))$ is formed in a polynomial form, which is fitted to the historic traffic flow data. Since in the mixed traffic flow both look-ahead controlled vehicles and conventional uncontrolled vehicles are traveling together, the rate of the autonomous vehicles κ and their energy-efficient parameter $R_{1,i}$ must be considered

$$\mathscr{F}(\rho_i(k), R_{1,i}, \kappa) = \sum_{j=1}^{n} c_j(R_{1,i}, \kappa)\rho_i(k)^j, \tag{8.3}$$

where the coefficients in the polynomials are formed as $c_j(R_{1,i}, \kappa) = \sum_{l=1}^{m} \left(d_l R_{1,i}^l \kappa^l\right)$ with constant d_l values. The mixed traffic model using Eqs. 9.1 and (8.3) is as follows:

$$\rho_i(k+1) = \rho_i(k) + \frac{T}{L_i}[-\mathscr{F}(\rho_i(k), R_{1,i}, \kappa) + q_{i-1}(k) + r_i(k) - s_i(k)] \tag{8.4}$$

Since highways contain several ramps, it is also required to model the dynamics of the queue on the controlled gates. The length of the queues can be calculated through the following linear relationship, see Csikos et al. (2014):

$$l_i(k+1) = l_i(k) + T\left(r_{i,dem}(k) - r_i(k)\right), \tag{8.5}$$

where l_i in the units of vehicle denote the queue length, r_i is the control input [veh/h], and the demand is $r_{i,dem}$.

Based on the mixed traffic model and considering the effects of κ and $R_{1,i}$ parameters, the stability of the traffic system will be analyzed. The examination is based on the Sum-of-Squares (SOS) method, in which the polynomial characteristics of the fundamental diagram Eq. (8.3) can be incorporated by Németh et al. (2017a).

One of the purposes of the analysis is to calculate the maximum inflows r_i and q_{i-1} function of $R_{1,i}$ and κ, by which the stability of the traffic flow can be guaranteed.

Using Eq. (8.4), the state-space representation of the mixed traffic system is given in the following form:

$$x(k+1) = f(R_{1,i}(k), x(k)) + g_1 u_{max}(R_{1,i}(k), \kappa(k)) + g_2 d(k), \tag{8.6}$$

where $x(k) = \rho_i(k)$ is the state of the system, $f(R_{1,i}(k), x(k))$ is a matrix, which incorporates smooth polynomial functions and its initial value is $f(R_{1,i}, 0) = 0$. $u_{max}(R_{1,i}(k), \kappa(k))$ is the function of the maximum controlled inflow $r_i(k)$ and $d_i(k) = q_{i-1}(k) - s_i(k)$ includes the measured disturbances of the system.

The stability analysis is based on the computation of the controlled invariant set using the Sum-of-Squares (SOS) programming method Tan and Packard (2008). The parameter-dependent control Lyapunov function is chosen in the following form:

$$\mathrm{V}(R_{1,i}(k), \kappa(k), x(k)) = V(x(k)) \cdot b(R_{1,i}(k), \kappa(k)), \tag{8.7}$$

where $b(R_{1,i}(k), \kappa(k))$ is an intuitively chosen parameter-dependent basis function. The existence of $\mathrm{V}(R_{1,i}(k), \kappa(k), x(k))$ is transformed into set emptiness conditions. Moreover, the domains of $R_{1,min} \leq R_{1,i}(k) \leq R_{1,max}$ and $\kappa_{min} \leq \kappa(k) \leq \kappa_{max}$ are also formulated in the set emptiness conditions. Using the generalized S-procedure Jarvis-Wloszek (2003), the set emptiness conditions can be transformed into the SOS existence problem, see Németh et al. (2017a).

As a result, an optimization problem is derived, in which the SOS conditions must be guaranteed. The optimization problem is to find a $u_{max}(R_{1,i}(k), \kappa(k))$ solution and feasible $\mathrm{V}(R_{1,i}(k), \kappa(k), x(k))$ for the following task:

$$\max \ u_{max}(R_{1,i}(k), \kappa(k)) \tag{8.8}$$

over $s_{1 \ldots 7} \in \Sigma_n;\ V(x(k)), b(R_{1,i}(k), \kappa(k)) \in \mathscr{R}_n$
such that

$$\begin{aligned}
&-\Bigg((V(f(R_{1,i}(k), \kappa(k), x(k)) + g u_{max}(R_{1,i}(k), \kappa(k)))- \\
&- V(x(k))) \cdot b(R_{1,i}(k), \kappa(k)) + \nu \cdot V(x(k))\Bigg)- \\
&- s_1\Bigg(V(x(k)) \cdot b(R_{1,i}(k), \kappa(k)) - (1-\varepsilon)\Bigg)- \\
&- s_2\Bigg(1 - V(x(k)) \cdot b(R_{1,i}(k), \kappa(k))\Bigg) - s_3 x(k)- \\
&- s_4\left(R_{1,i}(k) - R_{1,min}\right) - s_5\left(R_{1,max} - R_{1,i}(k)\right) - \\
&- s_6\left(\kappa(k) - \kappa_{min}\right) - s_7\left(\kappa_{max} - \kappa(k)\right) \in \Sigma_n,
\end{aligned} \tag{8.9}$$

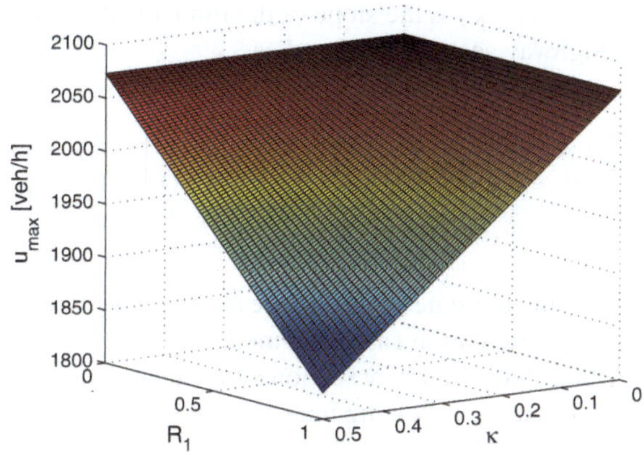

Fig. 8.1 Number of vehicles in the traffic flow based on the stability analysis

where the set of SOS polynomials in n variables is defined as $\Sigma_n := \{p \in \mathscr{R}_n \mid p = \sum_{i=1}^{t} f_i^2,\ f_i \in \mathscr{R}_n,\ i = 1, \dots, t\}$.

The result of the optimization Eq. (B.39) defines the maximum controlled invariant set, in which the system is stable with the function $u_{max}(R_{1,i}(k), \kappa(k))$. The computation method of the maximum controlled invariant set is found in Appendix B.

Thus, for the stability of the system the following inequality must be guaranteed:

$$q_{i-1}(k) + r_i(k) - s_i(k) \leq u_{max}(R_{1,i}(k), \kappa(k)) \tag{8.10}$$

The computation of the stability domain based on the results of VisSim simulations is demonstrated below Németh et al. (2017b). The maximum inflow u_{max} of the system depending on κ and $R_{1,i}$ is illustrated in Fig. 8.1. It is shown that increase in $R_{1,i}$ and κ reduce the maximum number of vehicles of the traffic flow u_{max}. Consequently, in the control design, it is necessary to find an appropriate coordination between the stability margin of the traffic system and energy-optimal cruising of the individual vehicles.

8.2 MPC-Based Coordinated Control Strategy

The control design of the highway is based on the model of the traffic flow Eq. (8.4) and the queue dynamics Eq. (8.5). In the control-oriented traffic model, the nonlinear dynamics Eq. (8.4) is linearized around $\rho_i = 0$. Consequently, the traffic model is valid to the maximum of the traffic density ρ_{max}. The mixed traffic model is as follows:

$$\rho_i(k+1) = \rho_i(k) + \frac{T}{L}\left(-\alpha(R_{1,i}, \kappa)\rho(k) + q(k) + r(k) - s(k)\right), \tag{8.11}$$

where $\alpha(R_{1,i},\kappa)$ is the slope of the fundamental diagram in $\rho_i = 0$. The equation of one highway section is formed as

$$\begin{bmatrix} \rho_i(k+1) \\ l_i(k+1) \end{bmatrix} = \begin{bmatrix} 1-\alpha(R_{1,i},\kappa) & 0 \\ 0 & 1 \end{bmatrix} \begin{bmatrix} \rho_i(k) \\ l_i(k) \end{bmatrix} + \begin{bmatrix} \frac{T}{L} & 0 \\ 0 & T \end{bmatrix} \begin{bmatrix} q(k)-s(k) \\ r_d \end{bmatrix} + \begin{bmatrix} \frac{T}{L} \\ -T \end{bmatrix} r_i(k), \tag{8.12}$$

By defining the state vector of one highway section $x_i(k) = \left[\rho_i(k)\ l_i(k)\right]^T$, the state-space representation can be formed.

Then, all the highway sections f are taken into consideration in order to compress the result in the following matrix form:

$$x(k+1) = A(R_{1,i}(k),\kappa(k))x(k) + B_1 w(k) + B_2 u(k), \tag{8.13}$$

where $x(k) = \left[x_1(k)\ x_2(k)\ \ldots\ x_f(k)\right]^T$ is the state vector, $w(k)$ is the disturbance, and $u(k)$ is the control. In the following, the $A(R_{1,i},\kappa)$ simplified denotion will be used instead of $A(R_{1,i}(k),\kappa(k))$.

In the traffic control problem, two main performances are defined.

- The total travel distance must be maximized in order to guarantee the maximum outflow of the highway section. In order to achieve this requirement, the maximum critical traffic density of the highway section $\rho_{crit,i}$ is introduced and the following tracking criterion is defined:

$$z_{1,i}(k) = \rho_i(k) - \rho_{crit,i}(k), \qquad |z_1| \rightarrow min. \tag{8.14}$$

  Note that $\rho_{crit,i}$ is the reference value in this performance specification.
- The length of the queue on the controlled ramp metering must be reduced to minimize the waiting time of the vehicles as given below:

$$z_{2,i}(k) = l_i(k), \qquad |z_2| \rightarrow min. \tag{8.15}$$

Since in the control design all of the highway sections are handled together, the performances of the sections are compressed to a vector as

$$z(k) = \left[z_{1,1}\ z_{2,1}\ z_{1,2}\ z_{2,2}\ \cdots\ z_{1,f}\ z_{2,f}\right]^T. \tag{8.16}$$

Since the traffic has relatively slow dynamics, the optimal selection of the current control input has high importance. The expected traffic dynamics or the expected changes in the traffic flow must be incorporated into the control design. Moreover, in the control design, constraints must be taken into consideration, e.g., the control inputs $r_i \geq 0$. Consequently, the model predictive control is applied to design the appropriate control interventions.

The MPC problem is described in a finite time horizon $n \cdot T$ ahead. The performance in the given horizon is calculated in the following way:

$$Z = \begin{bmatrix} C \\ CA \\ CA^2 \\ \vdots \\ CA^n \end{bmatrix} x(k) - \begin{bmatrix} z_{ref}(k) \\ z_{ref}(k+1) \\ \vdots \\ z_{ref}(k+n) \end{bmatrix} + \begin{bmatrix} 0 & 0 & \cdots & 0 \\ CB_1 & 0 & \cdots & 0 \\ CAB_1 & CB_1 & \cdots & 0 \\ \vdots & \vdots & \ddots & \vdots \\ CA^{n-1}B_1 & CA^{n-2}B_1 & \cdots & CB_1 \end{bmatrix} \begin{bmatrix} w(k) \\ w(k+1) \\ \vdots \\ w(k+n) \end{bmatrix} + \\ + \begin{bmatrix} 0 & 0 & \cdots & 0 \\ CB_2 & 0 & \cdots & 0 \\ CAB_2 & CB_2 & \cdots & 0 \\ \vdots & \vdots & \ddots & \vdots \\ CA^{n-1}B_2 & CA^{n-2}B_2 & \cdots & CB_2 \end{bmatrix} \begin{bmatrix} u(k) \\ u(k+1) \\ \vdots \\ u(k+n) \end{bmatrix} \tag{8.17}$$

Based on the reference values $\rho_{crit,i}(j)$ in Eq. (8.14), the reference signals $z_{ref}(j)$, $j \in \{k, k+n\}$ are defined. The performance in the given horizon in a compact form is as follows:

$$Z = \mathscr{C} - \mathscr{R} + \mathscr{B}_1 W + \mathscr{B}_2 U, \tag{8.18}$$

where $\mathscr{C}$ contains the current states of the system, $\mathscr{R}$ contains the reference values, W contains the disturbance values, and U contains the control input values. Note that both W and U also contain the forthcoming disturbances and control inputs, respectively.

In the general MPC design, the following cost function is minimized as follows:

$$J(U) = \frac{1}{2} \sum_{i=1}^{n} Z^T(U) Q Z(U) + U^T R U, \tag{8.19}$$

where Q and R are weighting matrices. Substituting Eq. (8.18) into the function Eq. (8.19), the cost function is transformed as

$$\begin{aligned} J(U) &= \frac{1}{2} U^T (\mathscr{B}_2^T Q \mathscr{B}_2 + R) U + (\mathscr{C}^T Q \mathscr{B}_2 + \mathscr{R}^T Q \mathscr{B}_2 + W^T \mathscr{B}_1 Q \mathscr{B}_2) U + \varepsilon \\ &= \frac{1}{2} U^T \phi U + \beta^T U + \varepsilon, \end{aligned} \tag{8.20}$$

where ε consists of all the constant components. Since ε is independent of the effect of U on $J(U)$, it can be omitted from the optimization problem.

Since $J(U)$ contains the forthcoming disturbances and only the current values are measured, the additional elements of W must be estimated. Several methods have been worked out to provide information about the future traffic flow, see e.g., Jones (2001). For example, a deep learning approach based on big data was presented in Lv et al. (2015) , statistics and neural networks were used in Moretti et al. (2015), while an adaptive Kalman filter approach was proposed by Guo et al. (2014).

The minimization of the cost function $J(U)$ also guarantees the performances Eq. (8.16). However, the cost function itself does not guarantee the stability of the system. Thus, constraints are built in the MPC optimization problem using the results of the SOS-based stability analysis. The SOS analysis results in an inequality Eq. (8.10), which must be guaranteed to provide stability.

Moreover, the states of the system $\rho_i(k)$ and $l_i(k)$ must be positive, which is a further constraint on the MPC problem given as

$$x(k) \geq 0 \tag{8.21}$$

for all $1 \leq k \leq n$ time steps.

Finally, from Eqs. (8.10), (8.20), and (8.21), the MPC control design problem is formed in the following way:

$$\min_{u(k)\ldots u(k+n)} \frac{1}{2} U^T \phi U + \beta^T U \tag{8.22}$$

such that

$$\begin{aligned} u_{max}(R_{1,i}(k), \kappa) &\geq q_{i-1}(k) + r_i(k) - s_i(k), \quad \forall i, k \\ x(k) &\geq 0, \quad \forall k \\ U &\in \mathbf{U}, \end{aligned} \tag{8.23}$$

where $\mathbf{U}$ contains the achievable control inputs. The MPC problem can be solved using standard quadratic programming methods, e.g. Gill et al. (1981), Schmid and Biegler (1994). The result of the computation Eq. (8.22) is a series of control inputs on the horizon $T \cdot n$. The control inputs are computed online during the cruising of the vehicle.

Since the entire traffic system the energy-efficient parameter R_1 has an important role, its intervention possibility is also built into the control design. The selection of $R_{1,i}$ in each section is important not only in the force/energy requirement of the vehicles but also in the dynamics and stability of the traffic system, see Eqs. (8.4) and (8.10). This parameter is incorporated in the MPC problem through the constraints, see Eq. (8.23). Figure 8.2 illustrates the effect of $R_{1,i}$ on the fundamental diagram. It shows the relationship between $R_{1,i}$ and the outflow of the traffic system $q_i(k)$. If $R_{1,i}$ is reduced, $q_i(k)$ increases. Since outflow $q_i(k)$ significantly influences $\rho_i(k)$ and $l_i(k)$, the parameter $R_{1,i}$ has effects on the state of the system $x_i(k)$.

In the coordinated control strategy with the MPC-based control design, the selection of $R_{1,i}$ is proposed. Considering the inflows $q_{i-1}(k)$, $r_i(k)$, $s_i(k)$, and the length of the queue of the controlled ramp metering $l_i(k)$, the energy-efficient parameter $R_{1,i}$ is computed for each section. Depending on the ramp metering, two different scenarios are distinguished as follows:

Fig. 8.2 Dependence of fundamental diagram on $R_{1,i}$ at $\kappa = 30\%$

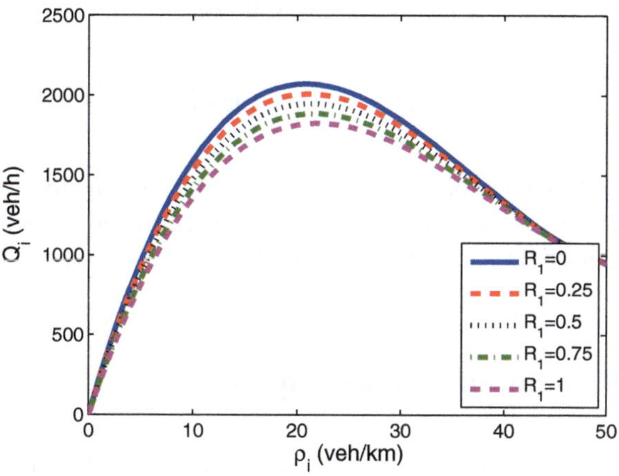

Fig. 8.3 Effect of $l_i(k)$ on $R_{1,i}(k)$

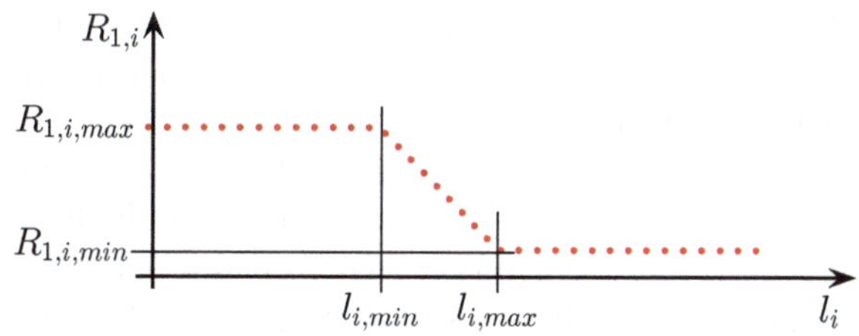

1. If the highway section i is controlled by the ramp metering r_i, then $R_{1,i}$ must depend on l_i. If l_i increases significantly, then $R_{1,i}$ must be reduced to guarantee larger traffic flow on this section. However, the increased flow results in increased r_i, by which the length of queue is reduced, see Eq. (8.5). Consequently, the parameter $R_{1,i}$ must be selected as the function of l_i according to Fig. 8.3. Here, $l_{i,min}$ and $l_{i,max}$ are design parameters.
2. If the highway section i does not include any controlled ramp metering $r_i(k)$, then $R_{1,i}$ must depend on the inflows and outflows. In this case, it is necessary to avoid the instability of the system while the maximum $R_{1,i}$ is selected. From the SOS programming method, the maximum traffic flow $u_{max}(R_{1,i}(k), \kappa(k))$ is calculated using Eq. (B.39). Exploiting the experience illustrated in Fig. 8.1, the maximum traffic flow can be expressed by the following form:

$$u_{max}(R_{1,i}(k), \kappa(k)) = u^0_{max} - u^1_{max} R_{1,i}(k)\kappa(k), \tag{8.24}$$

where u^0_{max} and u^1_{max} are selected constants, while $R_{1,i}(k)$ and $\kappa(k)$ are functions of k. Since $r_i(k) = 0$, in the upper limit of Eq. (8.10) can be transformed into the following form:

$$u^0_{max} - u^1_{max} R_{1,i}(k)\kappa(k) = q_{i-1}(k) - s_i(k), \tag{8.25}$$

and $R_{1,i}(k)$ is selected as follows:

$$R_{1,i}(k) = \min\left(1, \max\left(\frac{u_{max}^0 - q_{i-1}(k) + s_i(k)}{u_{max}^1 \kappa(k)}, 0\right)\right), \tag{8.26}$$

Based on the above scenarios, in the coordinated control strategy, first, the current $R_{1,i}$ values are computed for each section and second, the MPC problem Eq. (8.22) is solved to achieve the optimal control inputs $r_i(k)$.

8.3 Simulation Examples

Finally, the efficiency of the traffic control strategy is illustrated through a simulation example, which is performed in the high-fidelity microscopic traffic software VisSim. The purpose of the example is to show that the MPC-based coordinated strategy is able to control the highway ramps and the energy-efficient parameter of the vehicles guarantees the performances of the traffic system.

In the first simulation example, a 20-km-long section of the highway $M1$ between Budapest and Vienna is demonstrated. The highway section is divided into five segments, and it contains two controlled on-ramps and one off-ramp, see Fig. 8.4. During the simulation, it is necessary to minimize the lengths of the queues on the on-ramps, while the traffic flow and the energy saving of the vehicles are maximized.

The simulation parameter is $T = 30$ s sampling time in the prediction with $n = 12$ points, which leads to a total of 6 min prediction horizon. However, $T = 30$ s is too small sampling time for the intervention in the traffic control system, thus $T = 120$ s is selected with $n = 3$ points for the control horizon Frejo and Camacho (2012). Since the control input is computed as a flow value, it is transformed into green time with a 120 s cycle. Moreover, in the simulation $\kappa = 20\%$ value along the highway is considered. During the simulation, the signals in $w(k)$ are considered.

The results of the simulation are shown in Fig. 8.5. The simulation shows increasing traffic, whose maximum is approximately 3 h, see the density and the flow values at Fig. 8.5a, b. The critical density of the traffic is $\rho_{crit} = 25$ veh/km, whose tracking influences the ramps, see Fig. 8.5c. The efficiency of the prediction can be seen

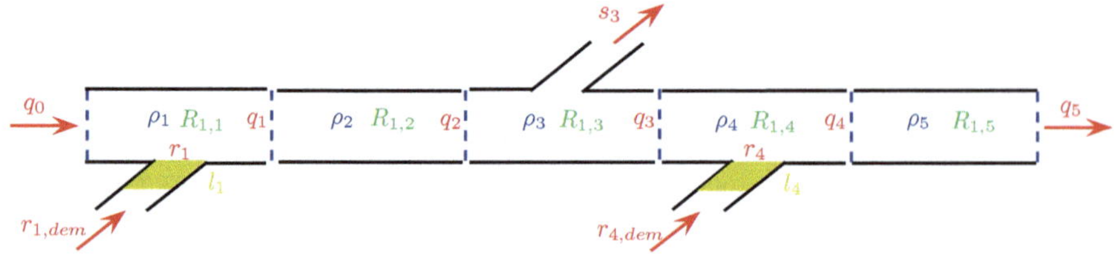

Fig. 8.4 Simulation scenario in the first example

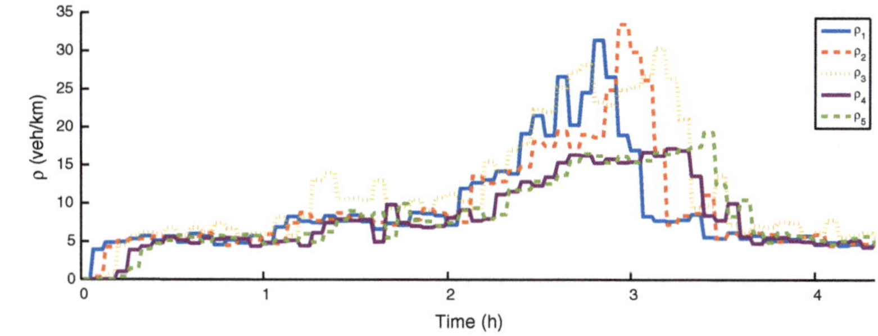

(a) Traffic density

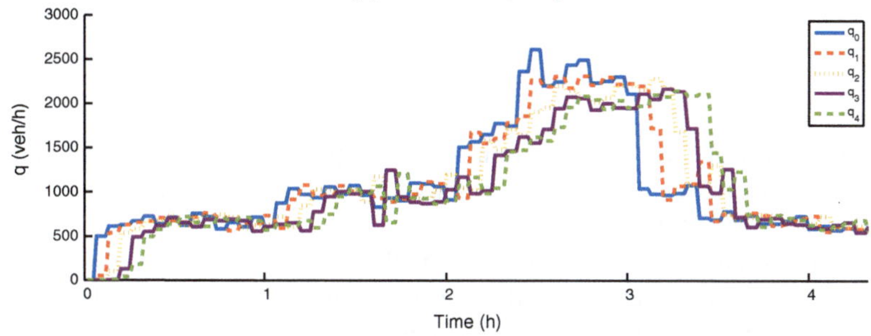

(b) q_{i-1} inflow of the sections

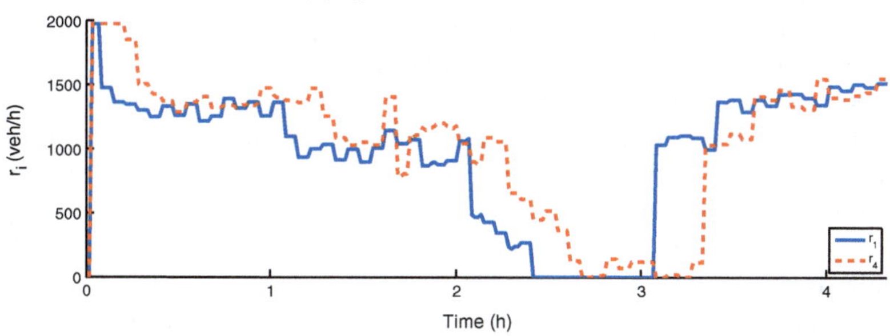

(c) Ramp control inflow

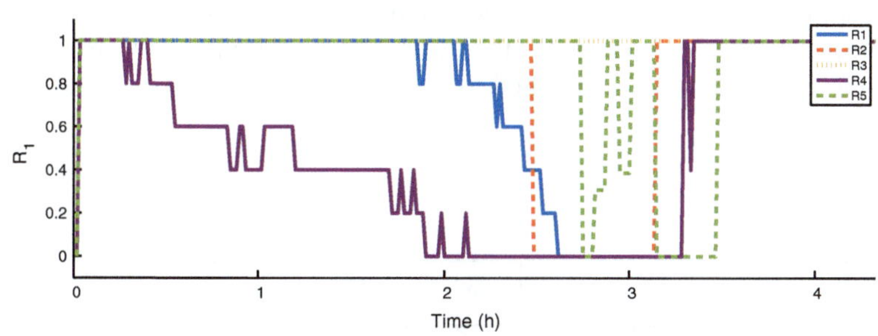

(d) Look-ahead parameter $R_{1,i}$

Fig. 8.5 Results of the VisSim simulation in the first example

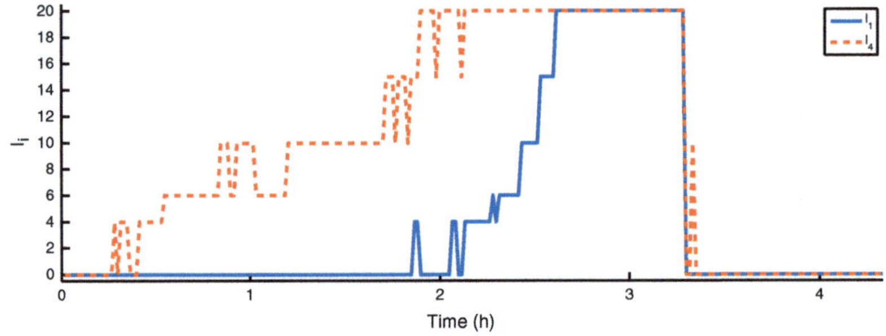

(e) Queue length on the controlled gates

Fig. 8.5 (continued)

in the dynamics of the r_i intervention. For example, at time 2.4 h, the value of r_1 significantly decreases, because q_0 increases in the future rapidly. However, ρ_1 is approximately 15 veh/km in $t = 2.4$ h.

The variation in economy parameters and the lengths of the queues are illustrated in Fig. 8.5d, e, respectively. The effect of the queue length on $R_{1,i}$ values is shown in the figures. When the queue length increases, the parameters $R_{i,1}$ decrease simultaneously, see the signals of l_1, l_4 and $R_{1,1}$, $R_{1,4}$. Moreover, $R_{1,2}$, $R_{1,3}$, $and R_{1,5}$ are selected to avoid the instability of the traffic and improve the flow capacity, e.g., at rush-hour traffic their values decrease, see between 2.6 . . . 3.5 h.

In the second simulation example, a 20-km-long section of the highway $M1$ between Budapest and Vienna is also demonstrated. The highway section is divided into five segments, and it contains two controlled on-ramps and one off-ramp, see Fig. 8.6. During the simulation, it is necessary to minimize the lengths of the queues on the on-ramps, while the traffic flow and the energy saving of the vehicles are maximized.

The simulation parameter is $T = 30$ s sampling time in the prediction with $n = 12$ points, which leads to a total of 6 min prediction horizon. However, $T = 30$ s is too small sampling time for the intervention in the traffic control system, thus $T = 120$ s is selected with $n = 3$ points for the control horizon Frejo and Camacho (2012). Since the control input is computed as a flow value, it is transformed into green

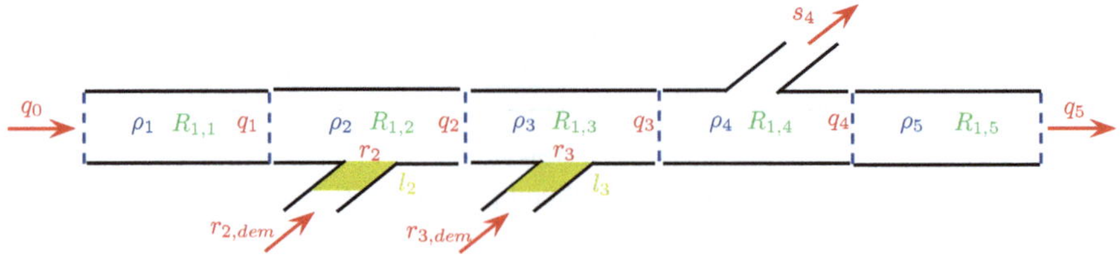

Fig. 8.6 Simulation scenario in the second example

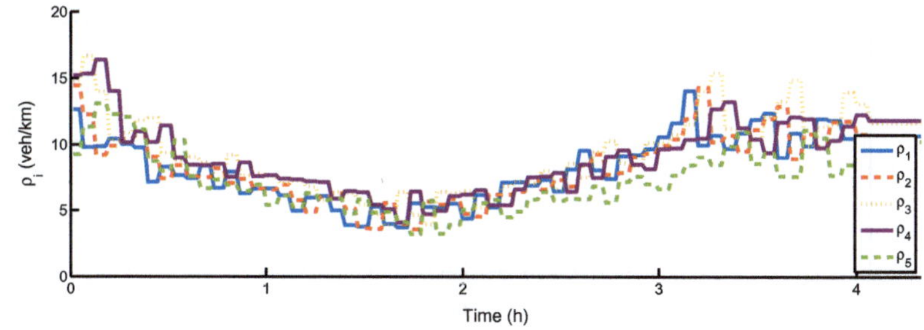

(a) Traffic density

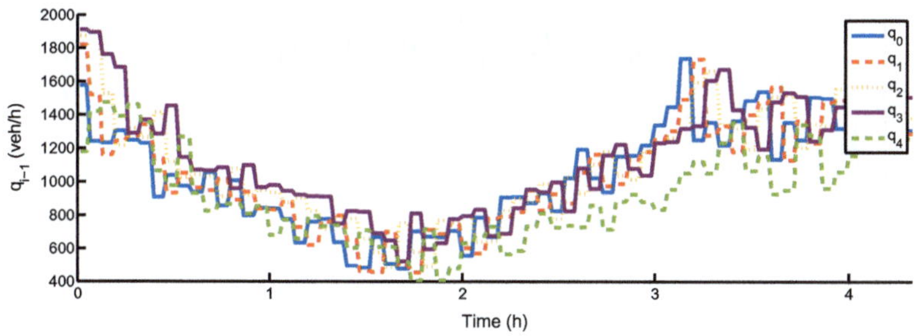

(b) q_{i-1} inflow of the sections

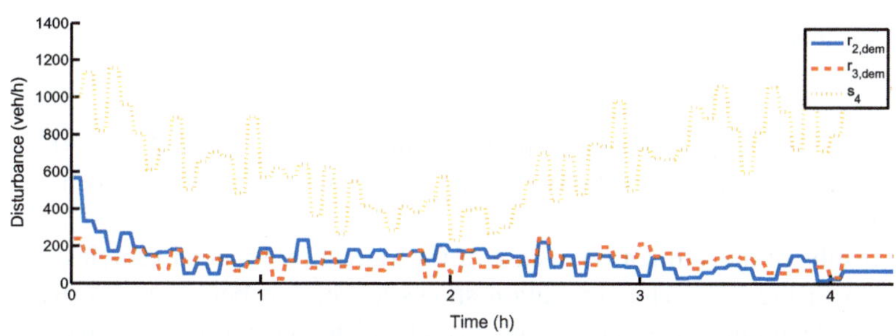

(c) Disturbance inflows of the sections

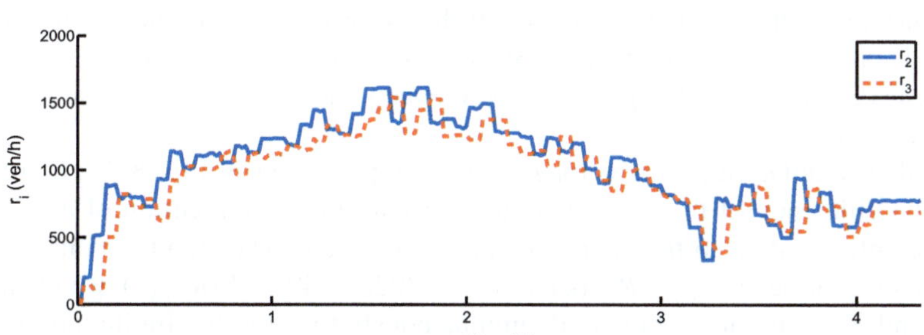

(d) Ramp control inflow

Fig. 8.7 Results of the VisSim simulation at $\kappa = 10\%$

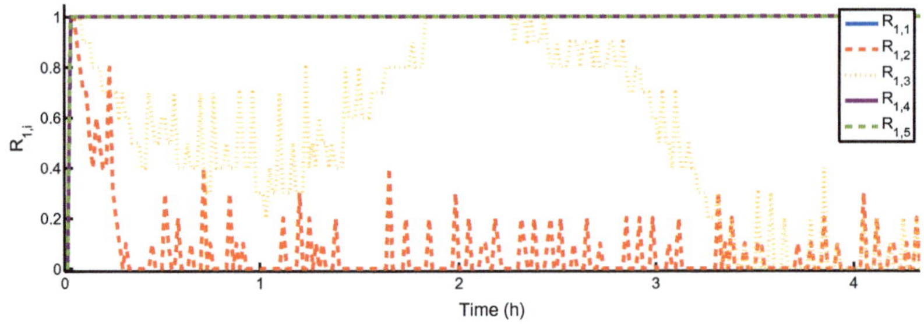

(e) Look-ahead parameter $R_{1,i}$

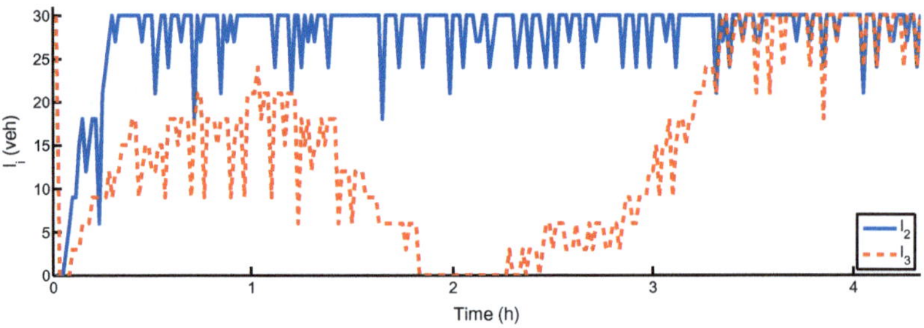

(f) Queue length on the controlled gates

Fig. 8.7 (continued)

time with a 120 s cycle. Moreover, in the simulation $\kappa = 10\%$ and $\kappa = 30\%$ values along the highway are considered. During the simulation, the signals in $w(k)$ are considered.

The results of the simulation in case of low rate of look-ahead vehicles $\kappa = 10\%$ are shown in Fig. 8.7. The simulation shows a scenario between two rush traffic peaks, whose minimum is approximately at 1.7 h, see the density and the flow values at Fig. 8.7a, b. Due to the reduction of the inflows q_0 and $r_{2,dem}$, $r_{3,dem}$, s_4, the controller is able to increase the controlled inflow r_i, while the stability of the system is preserved, see Fig. 8.7c, d. Since r_2, r_3 values are increased, the length of the queues in the controlled gates can be reduced and the look-ahead parameter is increased, see Fig. 8.7e, f.

Another simulation scenario for $\kappa = 30\%$ is presented in Fig. 8.8. The results show the efficiency of the control strategy, because the intervention of r_2, r_3, and $R_{1,i}$ are able to adapt to the change of the vehicle rate. The stability constraint of the system is formed as $u_{max}(R_{1,i}(k), \kappa(k)) = 2023 - 492\kappa(k)R_{1,i}(k)$, which means that κ and $R_{1,i}$ cannot be increased simultaneously to avoid the limitation of flow. The impact of the increase in κ is shown around the time 2h, where $R_{1,3}$ cannot be increased to the maximum value in $\kappa = 30\%$ scenario, while in $\kappa = 10\%$ scenario $R_{1,3} = 1$. Thus, the queue length is higher at $\kappa = 30\%$, as shown in Fig. 8.8f.

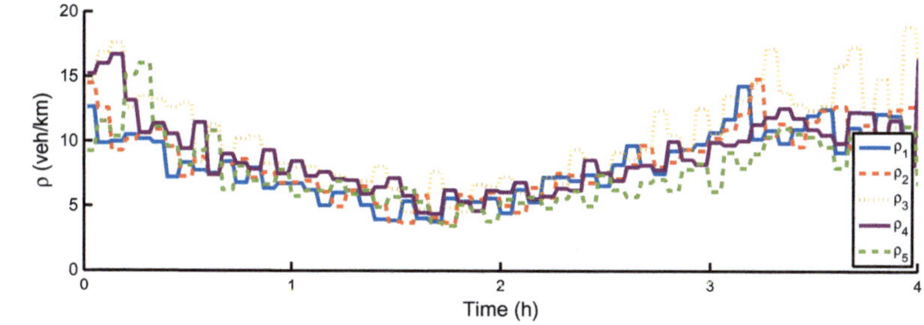

(a) Traffic density

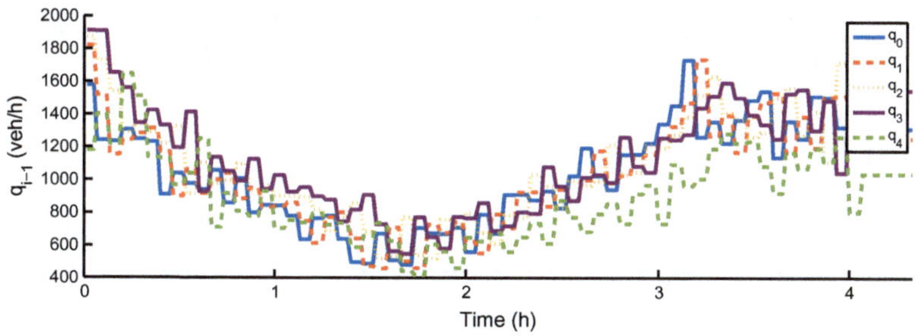

(b) q_{i-1} inflow of the sections

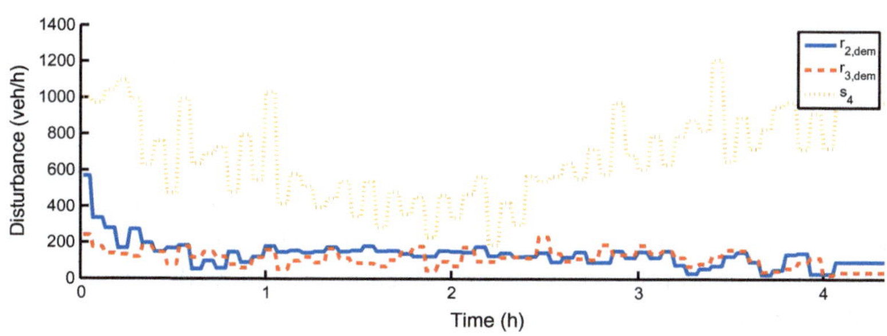

(c) Disturbance inflows of the sections

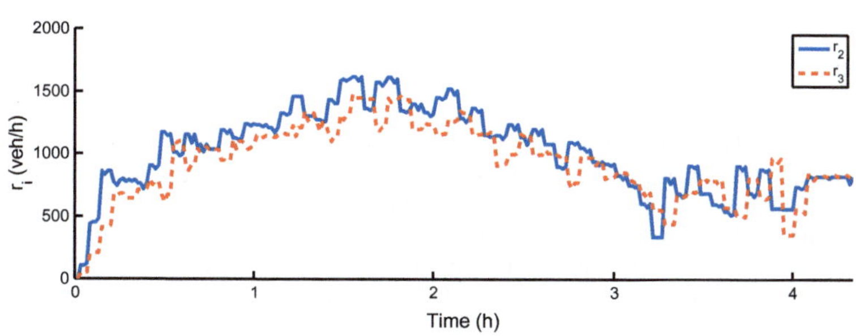

(d) Ramp control inflow

Fig. 8.8 Results of the VisSim simulation at $\kappa = 30\%$

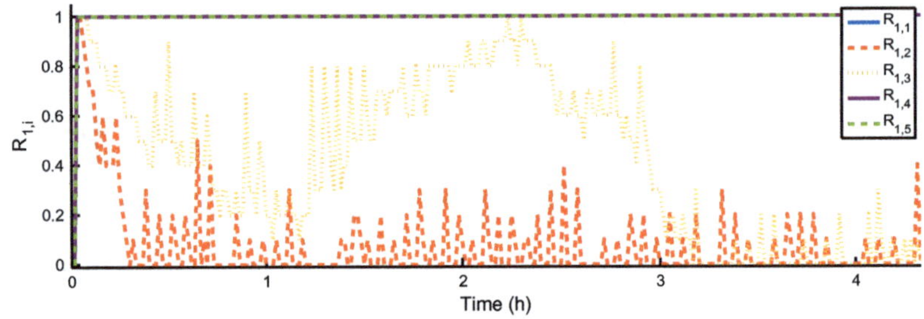

(e) Look-ahead parameter $R_{1,i}$

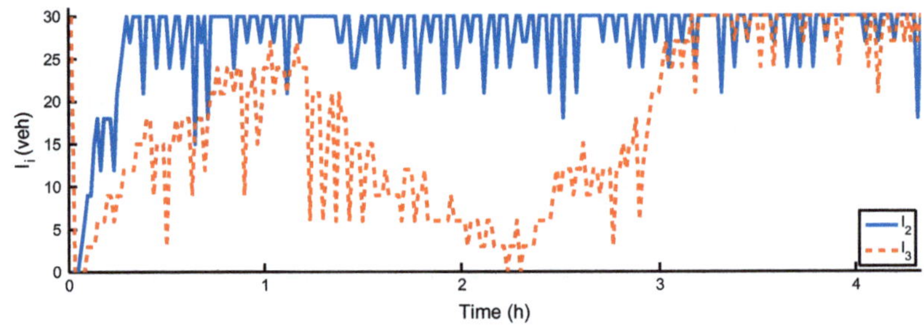

(f) Queue length on the controlled gates

Fig. 8.8 (continued)

Summarizing, the simulation examples show that the control strategy is able to guarantee the stability of the traffic flow and the control of the energy-efficient look-ahead vehicles. In the strategy, the inflow of the traffic system and the look-ahead parameter of the individual vehicles are coordinated. The control strategy is able to adapt to the variation in the traffic signals, such as the inflow disturbances and the ratio of the energy-efficient vehicles.

References

Ashton W (1966) The theory of traffic flow. Spottiswoode, Ballantyne and Co. Ltd., London

Csikos A, Tettamanti T, Varga I (2014) Nonlinear gating control for urban road traffic network using the network fundamental diagram. J Adv Transp 49(5):597–615

Frejo J, Camacho E (2012) Global versus local MPC algorithms in freeway traffic control with ramp metering and variable speed limits. IEEE Trans. Intell. Transp. Syst. 13(4):1556–1565

Gartner N, Wagner P (2008) Analysis of traffic flow characteristics on signalized arterials. Transp Res Rec 1883:94–100

Gill PE, Murray W, Wright M (1981) Practical optimization. Academic Press, London

Guo J, Huang W, Williams BM (2014) Adaptive kalman filter approach for stochastic short-term traffic flow rate prediction and uncertainty quantification. Transp Res Part C Emerg Technol 43(1):50–64

Jarvis-Wloszek Z (2003) Lyapunov based analysis and controller synthesis for polynomial systems using sum-of-squares optimization. PhD thesis, University of California, Berkeley
Jones WD (2001) Forecasting traffic flow. IEEE Spectr 38(1):90–91
Lv Y, Duan Y, Kang W, Li Z, Wang FY (2015) Traffic flow prediction with big data: a deep learning approach. IEEE Trans Intell Transp Syst 16(2):865–873
Messmer A, Papageorgiou M (1990) METANET-a macroscopic simulation program for motorway networks. Traffic Eng Control 31:466–470
Moretti F, Pizzuti S, Panzieri S, Annunziato M (2015) Urban traffic flow forecasting through statistical and neural network bagging ensemble hybrid modeling. Neurocomputing 167(1):3–7
Németh B, Bede Z, Gáspár P (2017a) Control design of traffic flow using look-ahead vehicles to increase energy efficiency. In: American control conference, pp 3530–3535
Németh B, Bede Z, Gáspár P (2017b) Modelling and analysis of mixed traffic flow with look-ahead controlled vehicles. In: 20th IFAC world congress
Papageorgiou M, Vigos G (2008) Relating time-occupancy measurements to space-occupancy and link vehicle-count. Transp Res Part C 16(1):1–17
Schmid C, Biegler L (1994) Quadratic programming methods for reduced hessian SQP. Comput Chem Eng 18(9):817–832
Tan W, Packard A (2008) Stability region analysis using polynomial and composite polynomial lyapunov functions and sum-of-squares programming. IEEE Trans Autom Control 53(2):565–571
Tettamanti T, Luspay T, Kulcsár B, Péni T, Varga I (2014) Robust control for urban road traffic networks. IEEE Trans. Intell. Transp. Syst. 15(1):385–398
Treiber M, Kesting A (2013) Traffic flow dynamics: data, models and simulation. Springer, Berlin

Chapter 9
Data-Driven Coordination Design of Traffic Control

Introduction and Motivation

The challenge posed by eco-cruise controlled vehicles is the modeling and control of the mixed traffic flow. Most of the novel traffic control design methods are based on the state-space representation of traffic flow dynamics. It is incorporated in several relationships (Messmer and Papageorgiou 1990), e.g., the conservation of vehicles, equilibrium speed equation, fundamental equation, and momentum equation. Although it can provide an enhanced description of traffic dynamics, due to the uncertainties, the estimation of model parameters may be difficult. For example, Frejo and Camacho (2012) proposed an identification method for traffic model parameters, especially the fundamental diagram, which has an important role in traffic control design. In a mixed traffic scenario, the identification problem can be more difficult, because the deviation of the measured data is more significant due to the varying speed profiles of the vehicles. The most important modeling approach for mixed traffic was summarized in the survey (Wageningen et al. 2015). The analysis of the traffic flow, in which semi-automated and automated vehicles were traveling together with conventional vehicles, was proposed by Schakel et al. (2010), van Arem et al. (2006). A control law which considers the different speed profiles of the semi-automated vehicles was proposed by Bose and Ioannou (2003), Zhang and Vahidi (2011). A fundamental diagram-based mixed traffic control design regarding the eco-cruise controlled vehicles was developed in Németh et al. (2015).

Since the control of the traffic flow based on the classical state-space representation for mixed traffic can be difficult due to the uncertainties, a data-driven approach is proposed in this chapter. Big data analysis has already been used in various traffic problems, but its application in the control of traffic flow dynamics is a novel field. Zhu et al. (2016) presented the idea of the path planning strategy of public vehicle systems, which used traffic data. Amini et al. (2017) proposed a comprehensive and flexible architecture based on a distributed computing platform for real-time traffic control. The architecture was based on systematic analysis of the requirements of

P. Gáspár and B. Németh, *Predictive Cruise Control for Road Vehicles Using Road and Traffic Information*, Advances in Industrial Control,
https://doi.org/10.1007/978-3-030-04116-8_9

the existing traffic control systems. Traffic flow prediction using a deep learning algorithm was presented in Lv et al. (2015). In that research, a deep architecture model was applied using autoencoders as building blocks to represent traffic flow features for prediction. Similarly, Lasso regression was used for traffic flow prediction in Li et al. (2015). Cell phone information-based big data analysis and control for transportation purposes were proposed in Dong et al. (2015), Zeyu et al. (2017).

The scope of this chapter is to analyze the problem of mixed traffic control with conventional and eco-cruise control vehicles. The Ordinary Least Squares (OLS) method for the examination of big data from the traffic flow is used. The goal of the analysis is to determine the traffic inflow of the network and the ratio of the eco-cruise controlled vehicles from the aspect of the traffic systems stability. In the following, data-driven coordinated traffic and vehicle control strategy are proposed, by which the inflow at the entrance gates and the speed profile of the eco-cruise controlled vehicles are influenced. Thus, the intervention possibilities are the green time of the traffic lights on the entrances and the speed profile of the eco-cruise controlled vehicles. The advantage of this approach is that in the proposed strategy, the fundamental diagram of the traffic dynamics, which contains several parameter uncertainties, is avoided. The resulting control strategy provides methodology for the handling of the mixed traffic scenarios with eco-cruise controlled and conventional vehicles. Moreover, in the method, the disturbances of the measured signals are also considered. The efficiency of the novel traffic and vehicle control approach is demonstrated through simulation scenarios on a freeway through the complex traffic simulator VisSim and the WEKA data mining software.

9.1 Architecture of the Proposed Traffic Control System

In this section, the architecture of the traffic control system and the modeling of the mixed traffic dynamics are proposed. Both of them are incorporated in the presence of the eco-cruise controlled vehicles in the traffic flow, as presented below.

The architecture scheme of the system is illustrated in Fig. 9.1. The traffic dynamics represents the traffic network, which is gridded into N number of segments. The

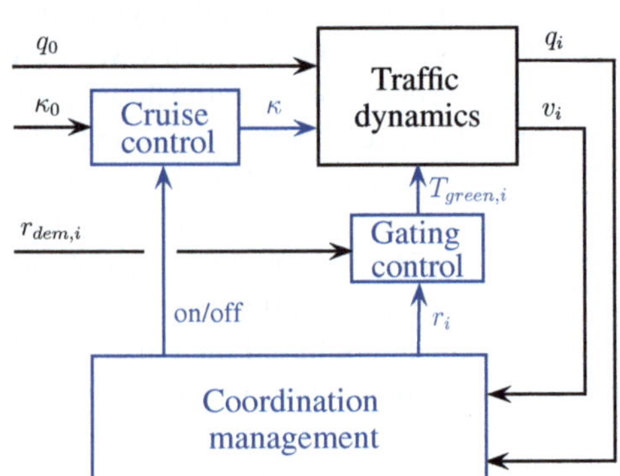

Fig. 9.1 Architecture scheme of the coordinated control

traffic flow of each segment is represented by a dynamical equation, which is based on the law of conservation. The relationship contains the sum of inflows and the outflows for a given segment i.

Traffic density ρ_i $[veh/km]$ is expressed in the following way:

$$\rho_i(k+1) = \rho_i(k) + \frac{T}{L_i}\left[q_{i-1}(k) - q_i(k) + r_i(k)\right], \tag{9.1}$$

where k denotes the index of the discrete time step, T is the discrete sample time, L_i is the length of the segment, q_i $[veh/h]$ and q_{i-1} $[veh/h]$ denote the inflow of the traffic in segments i, and $i-1$, r_i $[veh/h]$ is the sum of the controlled ramp inflow. Thus, an intervention possibility into the operation of the system is the green time of the traffic lights $T_{green,i}$ at the controlled gates (Papageorgiou and Vigos 2008), which are actuated by the gating control. The gating control is influenced by the coordination management, which decides about the entering vehicle inflow of the controlled gates r_i. It requires the measurements of the outflow q_i and the average traffic speed v_i of each segment. Thus, $r_i \leq r_{i,dem}$, where $r_{i,dem}$ is the entering vehicle flow demand at the controlled gates.

In the case of eco-cruise controlled automated vehicles, another intervention possibility into the dynamics of the traffic system is the modification of the speed selection strategy of the controlled vehicles individually. In this scheme, actuation means that the eco-cruise control functionality of the automated vehicles can be switched on or off. If the eco-cruise control functionality is switched on, the speed selection is energy efficient with significant variation. Otherwise, the speed of the vehicle is set at the speed limit, and the speed selection of the automated vehicle operates as in a conventional human-driven vehicle. κ_0 represents the ratio of the vehicles in the mixed traffic flow which has the ability to operate with the eco-cruise control strategy. However, the cruise control of some entering vehicles can decide to switch off this functionality. This decision is based on the coordination management of the cruise control and the gating traffic control. Consequently, the ratio of the vehicles with operating eco-cruise control strategy is κ.

The purpose of the coordination management is to guarantee the stability of traffic dynamics, i.e., the avoidance of congestion and maximize the average traffic speed. The coordination management level receives the q_i, v_i measurements, while it decides about r_i, κ. The roles of the gating control and the cruise control are to perform the decision on the lower level, such as traffic light actuation and vehicle speed computation and tracking.

The second important relation of the traffic dynamics is the fundamental relationship, which creates a connection between the outflow $q_i(k)$, traffic density $\rho_i(k)$, and the average traffic speed $v_i(k)$, see e.g., Ashton (1966). The fundamental relationship is formed as

$$q_i(k) = \rho_i(k)v_i(k). \tag{9.2}$$

Conventionally, the fundamental relationship is derived through historic measurements and depends on several factors, see Messmer and Papageorgiou (1990), Treiber and Kesting (2013). Thus, $v_i(k)$ and $q_i(k)$ can be formed in the traffic flow modeling studies as a nonlinear function of traffic density (Gartner and Wagner 2008), such as

$$v_i(k) = \mathscr{F}_v(\rho_i(k)), \tag{9.3}$$

$$q_i(k) = \mathscr{F}_q(\rho_i(k)), \tag{9.4}$$

where $\mathscr{F}_v$, $\mathscr{F}_q$ are nonlinear functions. In Sect. 4, the effect of κ on the average speed of the traffic is illustrated. In the modeling of the traffic flow dynamics, the effect of κ on $v_i(k)$ can be considered in the nonlinear function of $\mathscr{F}$ Németh et al. (2015), with the extension of $\mathscr{F}_v$, such as

$$v_i(k) = \mathscr{F}(\rho_i(k), \kappa(k)). \tag{9.5}$$

Thus, traffic density can be computed through the fundamental relationship (9.2) and (9.5) as given below:

$$\rho_i(k) = \frac{q_i(k)}{\mathscr{F}(\rho_i(k), \kappa(k))}. \tag{9.6}$$

The determination of the function $\mathscr{F}(\rho_i(k), \kappa(k))$ is based on historical measurements (Gartner and Wagner 2008), which is evaluated through statistical analysis and identification methods. However, the deviation of the measurements can be significant. It results in uncertainty in the traffic flow model, which is a difficulty for the traffic control design. Therefore, in the following, the design of the coordination and the control are based on a data-driven approach.

9.2 Optimal Coordination Strategy Based on Traffic Flow Data

In this section, the method of the data-driven analysis for the modeling of the traffic flow dynamics is proposed. The analysis is based on the measurements of traffic flow data, which are the average speed of the traffic v_i, the output traffic flow q_i, and the density ρ_i in all sections i. The goal of the formulation is to find a control-oriented linear model by which the traffic flow dynamics can be approximated. In the modeling, the current data in time k and previous information $k - j$, $j > 0$ are considered. First, the fundamentals of the LS method are presented, and second, it is applied to current modeling problems.

9.2.1 Fundamentals of the LS Method

In this research, the conventional LS method combined with subset selection is used to generate prediction models for the traffic flow. In the following, the most important features are briefly summarized. A detailed description is found in Wang and Witten (1999).

Consider a dataset with n independent instances, k input variables, and one output variable. The instances are written in the form of an $n \times k$ design matrix X. Let ζ^* be the parameter vector of the true model. Through the application of X and ζ^*, the resulting output vector y is determined as

$$y = X\zeta^* + \varepsilon, \tag{9.7}$$

where ε is the noise vector, whose elements have normal distribution $N(0, \sigma^2)$, with σ variance. It is assumed that σ^2 is known or it can be estimated, which is denoted by $\hat{\sigma}^2$. $\mathscr{M}(\zeta)$ denotes an estimated model that has a unique parameter vector ζ while the true model is denoted by $\mathscr{M}(\zeta^*)$.

The purpose of the modeling task is to find a model from the entire model space $\mathbb{M} = \{\mathscr{M}(\zeta) : \zeta \in \mathbb{R}^k\}$ whose accuracy is the greatest on the given dataset. The models can be produced by numerous algorithms, e.g., the LS method, the LS subset selection, shrinkage, RIC, and CIC methods, see Wang and Witten (1999). These methods can reduce the dimension of the models by discarding the redundant variables.

In order to evaluate a model, the distance $\mathscr{D}$ between the candidate model and the true model must be known. The distance is calculated as

$$\mathscr{D}(\mathscr{M}(\zeta^*), \mathscr{M}(\hat{\zeta})) = \frac{||y - \hat{y}||^2}{\sigma^2}, \tag{9.8}$$

where $|| \cdot ||$ denotes the $\mathscr{L}_2$ norm, y is the measured output, $\hat{y}$ is the estimated output, and σ^2 is the variance of the noise. The final task is to create a model which minimizes the following expression:

$$\min_{\zeta} \mathscr{D}(\mathscr{M}(\zeta^*), \mathscr{M}(\zeta)) \tag{9.9}$$

In the LS regression, the estimated output is expressed in the following way $\hat{y} = X\hat{\zeta}$, where $\hat{\zeta}$ is the parameter vector of the model. The distance is reformulated as

$$\mathscr{D}(\mathscr{M}(\zeta^*), \mathscr{M}(\hat{\zeta})) = \frac{||y - X\hat{\zeta}||^2}{\sigma^2}. \tag{9.10}$$

The solution of the optimization task can be obtained by a partial derivation according to $\hat{\zeta}$. The solution is the following: $\hat{\zeta} = (X^T X)^{-1} X^T y$. The estimated output $\hat{y}$ can be expressed using the orthogonal projection matrix $\mathscr{P} = X(X^T X)^{-1} X^T$ such as $\hat{y} = \mathscr{P} y$.

In the big-data-based method, the distance can be expressed with the following form of the training sets:

$$\frac{(y - X\hat{\zeta})^2}{\sigma^2} = \sum_{i=1}^{\omega} \frac{||y_i - X_i\hat{\zeta}||^2}{\sigma^2}, \tag{9.11}$$

where y_i and X_i are one of the training sets, while ω is the number of the training sets.

The difficulty in the model estimation using the big-data-based method is that the estimated model should be obtained from a large number of measurements efficiently in ζ, i.e., within a short time. If k increases, the number of the subset models increases by 2^k. It may lead to a computationally unfeasible task. Therefore, the efficient solution of the optimization task (9.9) requires the determination of the preferences among the sets of variables in ζ. The subset selection is based on the instances X, whose subset models are generated. In the generation of subset models, it is necessary to reduce the complexity of the big data analysis, while information loss from the data must be negligible, Wang and Witten (2002).

Using the subset selection procedure, the increasing k generates $k+1$ subset models. Thus, in the big data analysis, ranking algorithms are applied, which support the determination of the best order of the variables (Shibata 1981, Thompson 1978).

9.2.2 Modeling the Traffic Flow Dynamics

Through the LS method, the function of the traffic system outflow $q_N(k)$ is approximated through $\hat{q}_N(k)$. In the regression analysis, the next form is selected as

$$\begin{aligned}\hat{q}_N(k+1) = \delta_{q_N} + \sum_{l=k-j}^{k} \Bigg(\sum_{i=0}^{N} a_{i,l} \Big(q_i(l) + r_i(l) \Big) + \\ + \sum_{i=1}^{N} b_{i,l} v_i(l) + \sum_{i=1}^{N} c_{i,l} \rho_i(l) + d_l \kappa(l) \Bigg),\end{aligned} \tag{9.12}$$

where j is a design parameter, which represents the past data. In (9.12), the following data are incorporated:

- $q_i(l) + r_i(l)$ are the inflows of each segment, in which r_i represents the inflows on the controlled ramps, considering the current k and the past j information,
- $v_i(l)$ is the average traffic speed on each segment, considering the current k and the past j information,
- $\rho_i(l)$ is the traffic density on each segment, considering the current k and the past j information,
- κ is the ratio of the eco-cruise controlled vehicles, considering the current k and the past j information.

The parameters $\delta_{q_N}, a_{i,l}, b_{i,l}, c_{i,l}, and d_l$ are determined with the LS algorithm through the big data analysis.

The result of the regression is a linear function of $\hat{q}_N(k+1)$ in the form of (9.12), which approximates $q_N(k+1)$, such as

$$q_N(k+1) = \hat{q}_N(k+1) + e_{q_N}(k+1), \tag{9.13}$$

where $e_{q_N}(k+1)$ is the error of the prediction. Since $e_{q_N}(k+1)$ is unknown, in the following, it is handled as a disturbance of the system.

The linear regression form of the system with $e_{q_N}(k+1)$ can bc transformed into a state-space representation, such as

$$q_N(k+1) = \delta_{q_N}(k) + A_{q_N}x(k) + B_{q_N}u(k) + D_{q_N}w(k), \tag{9.14}$$

where x contains the states of the system, which can be measured currently or at the past of the process. Thus, most of the components in the regression (9.12) are incorporated in x, such as

$$x(k) = \begin{bmatrix} [q_1(k-j)\dots q_N(k)]^T \\ [v_1(k-j)\dots v_N(k)]^T \\ [\rho_1(k-j)\dots \rho_N(k)]^T \\ [r_1(k-j)\dots r_N(k-1)]^T \\ [\kappa(k-j)\dots \kappa(k-1)]^T \\ [q_0(k-j)\dots q_0(k-1)]^T \end{bmatrix}, \tag{9.15}$$

where the size of $x(k)$ is $M \times 1$ and M is the number of the states. Furthermore, the control inputs of the system are the signals, by which the traffic system (9.12) in k can be influenced, such as $u^T = \left[r_1(k)\dots r_N(k)\ \kappa(k)\right]$, and contains unknown signals of (9.12) in k and the prediction error in $k+1$: $w(k) = \left[q_0(k)\ e_{q_N}(k+1)\right]$. Thus, the control inputs of the system are the controlled gates and the ratio of the eco-cruise controlled vehicles, while the physical disturbance is the uncontrolled inflow into the traffic network.

Although (9.14) provides accurate information about the outflow of the system, the relationship between $u(k)$, $w(k)$, and $q_N(k+1)$ is not identified sufficiently. The reason for it is the state vector $x(k)$, whose elements have significant role in the value of the outflow. However, (9.14) does not provide information about the impact of $u(k)$, $w(k)$ on $x(k)$. Thus, it is necessary to expand the traffic flow model with further relationships, such as

$$x_i(k+1) = \delta_i + A_i x(k) + B_i u(k) + D_i w(k), \tag{9.16}$$

where $x_i(k)$ represents the ith state in $x(k)$, where $i = 1\dots M$. In this equation, δ_i, A_i, B_i, D_i are parameter vectors, which are determined through the LS method based on the traffic data. The relationships (9.14), (9.16) result in the control-oriented state-space representation of the system as

$$\begin{bmatrix} q_N(k+1) \\ x(k+1) \end{bmatrix} = \delta + Ax(k) + Bu(k) + Dw(k), \tag{9.17}$$

where the vector δ and the matrices A, B, D are yielded by δ_{q_N}, δ_i and A_{q_N}, B_{q_N}, D_{q_N}, A_i, B_i, D_i. Note that if $q_N(k)$ is in $x(k)$, then (9.17) is reduced to

$$x(k+1) = \delta + Ax(k)+Bu(k) + Dw(k). \tag{9.18}$$

The resulting state-space formulation of the system provides an efficient representation of the traffic flow dynamics, which can be used for control design purposes.

9.3 Optimal Coordination Strategy Based on Minimax Method

The goal of this section is to find an optimal control method by which the eco-cruise vehicles and the controlled inflow of the traffic gates are coordinated. The aim of the entire coordination is to guarantee the maximum outflow of the traffic network q_N through the ratio of the eco-cruise controlled vehicles κ_0 and the inflow on the controlled ramp r_i. Moreover, the control-oriented representation of the system (9.17) contains disturbances, whose impact on q_N must be minimized. It leads to a minimax optimization problem, which is introduced in the following.

In the coordination of the traffic system and autonomous vehicles various performances must be guaranteed.

1. It is necessary to reach the maximum outflow of the traffic network $q_N(k+1)$, which is at the critical density $\rho_{N,crit}$ of the system (Daganzo 1997). If $\rho_N(k+1) > \rho_{N,crit}$, then it can result in congestion and significant reduction in $q_N(k+1)$. Since the data-based linear approximation of the traffic flow dynamics (9.17) does not have information about the nonlinear characteristics of the process, the performance is defined as a tracking task, such as

$$z_1 = q_N(k+1) - q_{N,max}, \qquad |z_1| \rightarrow min. \tag{9.19}$$

   The values of the critical density and the maximum outflow $q_{N,max}$ are determined through the provisional analysis of the traffic network.
2. Another performance of the system is related to the controlled vehicles. Since the eco-cruise control strategy minimizes the energy consumption of the vehicles (Németh and Gáspár 2017), it is required to maximize $\kappa(k)$. Since $\kappa(k)$ is a ratio, whose value ranges between $[0 \ldots 1]$, the performance maximization of $\kappa(k)$ can be expressed as a minimization criterion

$$z_2 = 1 - \kappa(k), \qquad |z_2| \rightarrow min, \tag{9.20}$$

   where $1 - \kappa(k)$ represents the ratio of the uncontrolled vehicles. The value of $\kappa(k)$ can be expressed from (9.17) using

$$\kappa = D_\kappa u, \tag{9.21}$$

where $D_\kappa = [0 \;\; 1]$.

The performances of the system are compressed into a performance vector

$$z = \begin{bmatrix} z_1 \\ z_2 \end{bmatrix}. \tag{9.22}$$

Moreover, in the coordination process some limits of the control inputs must be incorporated.

1. The control input $r_i(k), i = 1 \dots N$ is a positive variable and it has physical limits. Moreover, it is necessary to limit the variation of $r_i(k)$ to avoid the rapid actuation change. Therefore, two constraints on $r_i(k)$ are defined, such as

$$r_i(k) \geq \max(0, r_i(k-1) - \Delta r_i), \tag{9.23}$$
$$r_i(k) \leq \min(r_{i,max}, r_i(k-1) + \Delta r_i), \tag{9.24}$$

   where Δr_i is the maximum of $r_i(k)$ variation and $r_{i,max}$ is the maximum of $r_i(k)$.
2. Another control input κ also has limits. The lower limit is $\kappa(k) \geq 0$ and the ratio of the controlled vehicle can have a maximum κ_{max}. κ_{max} depends on the ratio of the automated vehicles in which the speed can be controlled given as

$$\kappa(k) \geq 0, \tag{9.25}$$
$$\kappa(k) \leq \kappa_{max}. \tag{9.26}$$

The goal of the coordination strategy is to guarantee the performances z_1, z_2 of the system, while the constraints are kept. The coordination problem can be solved efficiently through the minimization of the objective function $\frac{1}{2}z^T Q z$, where

$$Q = \begin{bmatrix} Q_{z_1} & 0 \\ 0 & Q_{z_2} \end{bmatrix} \tag{9.27}$$

is a weighting matrix, which provides a balance between z_1 and z_2. Thus, the optimization problem is

$$\min_{\kappa(k), r_i(k)} \frac{1}{2} z^T Q z \tag{9.28}$$

such that

$$r_i(k) \geq \max(0, r_i(k-1) - \Delta r_i), \tag{9.29}$$
$$r_i(k) \leq \min(r_{i,max}, r_i(k-1) + \Delta r_i), \tag{9.30}$$
$$\kappa(k) \geq 0, \tag{9.31}$$
$$\kappa(k) \leq \kappa_{max}. \tag{9.32}$$

The quadratic function in (9.28) is expressed using (9.17), (9.19), (9.20), and (9.21) as

$$\begin{aligned}\frac{1}{2}z^T Qz &= \frac{1}{2}\gamma^T Q_{z_1}\gamma + \frac{1}{2}(1 - u^T(k)D_\kappa^T)Q_{z_2}(1 - D_\kappa u(k)) \\ &= \mathcal{Q}_1 + \mathcal{Q}_2 + \mathcal{Q}_3,\end{aligned} \tag{9.33}$$

with the notation $\gamma = \delta(k) + Ax(k) + Bu(k) + Dw(k) - q_{N,max}$. The components of the quadratic function are the following:

$$\begin{aligned}\mathcal{Q}_1 &= \frac{1}{2}(u^T(k)(B^T Q_{z_1} B + D_\kappa^T Q_{z_2} D_\kappa)u(k)) + \\ &\quad + w^T(k)D^T Q_{z_1} Dw(k),\end{aligned} \tag{9.34}$$

$$\begin{aligned}\mathcal{Q}_2 &= (\delta(k) + Ax(k))^T(Bu(k) + Dw(k)) + \\ &\quad + u^T(k)B^T Q_{z_1} Dw(k) - D_\kappa u(k),\end{aligned} \tag{9.35}$$

$$\mathcal{Q}_3 = \frac{1}{2}(\delta(k) + Ax(k))^T Q_{z_1}(\delta(k) + Ax(k)) + \frac{1}{2}Q_{z_2}. \tag{9.36}$$

The result of (9.33) shows that the value of $\mathcal{Q}_3$ contains only measurements through $\delta(k)$ and $x(k)$. Thus, it is a constant term in the optimization, which means that it can be ignored during the computation of $u(k)$. Moreover, in $\mathcal{Q}_1$, $\mathcal{Q}_2$, the control input $u(k)$ and the disturbance signal are mixed together. However, the value of $w(k)$ is unknown during the optimization problem (9.28). Therefore, in the following, the minimization problem (9.33) is transformed into a minimax optimization problem. In the minimax optimization, it is considered the worst-case scenario in terms of disturbance. It means that the purpose of $w(k)$ is to maximize $z^T Qz$, which is the worst case during the process. Thus, it is necessary to find $u(k)$, in whose computation the worst case is considered.

First, the sum of $\mathcal{Q}_1 + \mathcal{Q}_2$ from (9.33) is separated in the following way:

$$\begin{aligned}J_u &= \frac{1}{2}(u^T(k)(B^T Q_{z_1} B + D_\kappa^T Q_{z_2} D_\kappa)u(k)) + \\ &\quad + ((\delta(k) + Ax(k))^T B + \frac{1}{2}w^T(k)D^T Q_{z_1} B)u(k),\end{aligned} \tag{9.37}$$

$$\begin{aligned}J_w &= \frac{1}{2}w^T(k)D^T Q_{z_1} Dw(k) + \\ &\quad + ((\delta(k) + Ax(k))^T D + \frac{1}{2}u^T(k)B^T Q_{z_1} D)w(k).\end{aligned} \tag{9.38}$$

Second, the following new optimization problems are formed:

$$\min_{\kappa(k), r_i(k)} J_u \tag{9.39}$$

such that

$$r_i(k) \geq \max(0, r_i(k-1) - \Delta r_i), \tag{9.40}$$
$$r_i(k) \leq \min(r_{i,max}, r_i(k-1) + \Delta r_i), \tag{9.41}$$
$$\kappa(k) \geq 0, \tag{9.42}$$
$$\kappa(k) \leq \kappa_{max}, \tag{9.43}$$

and the next optimization is related to the worst-case effect, in which $w(k)$ is handled as an optimization variable

$$\max_{q_0(k), e(k+1)} J_w = \min_{q_0(k), e(k+1)} -J_w \tag{9.44}$$

such that

$$q_0(k) \leq q_{0,lim}, \tag{9.45}$$
$$|e(k+1)| \leq e_{lim}, \tag{9.46}$$

where $q_{0,lim}$, e_{lim} are the limits of the uncontrolled inflow and the error of the function approximation, respectively.

Since both optimization problems, (9.39) and (9.44), are transformed into constrained minimization problems, the two tasks are merged into one constrained minimization problem. During the merge process, the costs J_u and $-J_w$ are scaled with the weights Q_u and Q_w, which represent the balance between the costs. For example, if Q_u has a high value, then (9.39) has priority over (9.44). This example represents that the LS method has a small error and the impact of $q_0(k)$ is low on $q_N(k+1)$. Thus, the importance of the outflow maximization is higher than the consideration of the disturbance impact.

Then, the third step is the formulation of the final optimization problem given as

$$\min_{\kappa(k), r_i(k), q_0(k), e(k+1)} Q_u Ju - Q_w J_w =$$
$$\min_{\kappa(k), r_i(k), q_0(k), e(k+1)} \frac{1}{2} U^T \Phi U + \beta U \tag{9.47}$$

such that

$$r_i(k) \geq \max(0, r_i(k-1) - \Delta r_i), \tag{9.48}$$
$$r_i(k) \leq \min(r_{i,max}, r_i(k-1) + \Delta r_i), \tag{9.49}$$
$$\kappa(k) \geq 0, \tag{9.50}$$

$$\kappa(k) \leq \kappa_{max}, \tag{9.51}$$

$$q_0(k) \leq q_{0,lim}, \tag{9.52}$$

$$|e(k+1)| \leq e_{lim}, \tag{9.53}$$

where $U = \left[\kappa(k)\ r_i(k)\ q_0(k)\ e(k+1)\right]^T$ and

$$\Phi = \begin{bmatrix} -Q_u(B^T Q_{z_1} B + D_\kappa^T Q_{z_2} D_\kappa) & Q_w B Q_{z_1} D \\ -Q_u D^T Q_{z_1} B & Q_w D^T Q_{z_1} D \end{bmatrix} \tag{9.54}$$

$$\beta = \left[-Q_u(\delta(k) + Ax(k))^T B\ Q_w(\delta(k) + Ax(k))^T D\right]. \tag{9.55}$$

The results of the optimization are the intervention signals $\kappa(k)$, $r_i(k)$, which are computed through the coordination management. The solution of the task (9.47) is based on an optimization algorithm, which is able to handle the nonlinear constraints, see e.g. Gill et al. (1981), Coleman and Li (1996).

Finally, in the fourth step, it is necessary to actuate with the computed $r_i(k)$ and $\kappa(k)$. In the case of the controlled gates, it is necessary to guarantee the limited number of inflow vehicles based on $r_i(k)$. Moreover, in the case of the computed κ, it necessary to select the speed profile of automated vehicles which enter the network. Their speed profile can be conventional with the maximum speed limit or eco-cruise controlled with energy-optimal speed profile, as detailed above. In the coordination strategy, it is assumed that the decision cannot be modified along the route of the automated vehicle in the controlled network for safety reasons. Thus, the type of the speed profile is decided at the entrance of the network. Consequently, it is necessary to determine the ratio of the entering eco-cruise controlled vehicles κ_{in} by which $\kappa(k)$ can be reached.

For the computation of κ_{in}, the dynamics of κ variation is modeled via a transfer function $H(z)$ with the input $\kappa_{in}(k)$ and output $\kappa(k)$ as

$$H(z) = \frac{\beta_2 z^2 + \beta_1 z + \beta_0}{z^2 - (\alpha + 1)z + \alpha}, \tag{9.56}$$

where β_2, β_1, β_0, α are the parameters of the approximating system. Since it is necessary to guarantee $\kappa(k)$ for the traffic network (9.47), a PI controller is tuned to compute κ_{in} from the error between the actual and the required $\kappa(k)$ (Åström and Hägglund 2006). In the control system, κ_{in} is actuated on the entering automated vehicles.

9.4 Simulation Examples

The efficiency of the coordination strategy is demonstrated through the following simulation scenarios. In the example, a 7-km-long hilly section of the freeway $M1$ between Budapest and Vienna with two lanes is considered. The section is divided into $N = 5$ segments and it has a controlled ramp inflow r_1 in the first segment.

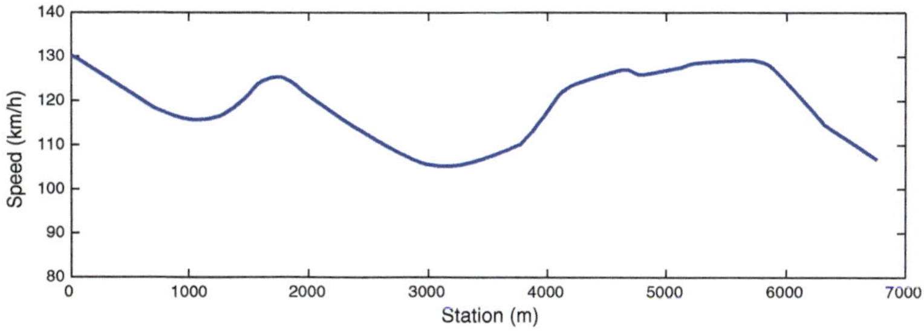

Fig. 9.2 Speed profile of the eco-cruise controlled vehicles

The traffic contains $\kappa_{max} = 50\%$ rate of eco-cruise controlled vehicles, whose speed profile is designed through (Németh and Gáspár 2017). Its form is found in Fig. 9.2, which significantly differs from the speed profile of the conventional vehicles with constant 130 km/h.

The collection of the data was carried out with the VisSim traffic simulator. It performed more than 600 mixed traffic simulations with various $\kappa_0 = 0; 10; 20 \ldots 50\%$ and $q_0 = 750; 1000; 1250 \ldots 5000$ veh/h average inflow values, while the vehicles arrived randomly in the network at all entrances.

Based on the results of a large number of traffic simulations using the VisSim simulator, the LS big data analysis was performed. In the analysis, the data mining WEKA software was used, in which the pace regression algorithm was implemented (Witten and Frank 2005). The advantage of the method is that it requires little online computation, while the complex operations are solved offline. The analysis has yielded the linear regression of $q_5(k+1)$ in the form of (9.12). Throughout the analysis, the $j = 1$ past data were considered with 5 min horizon backward. During the big data analysis, the best first subset selector algorithm was used to preprocess the training set for the pace regression algorithm. The subset selection resulted in 10 relevant attributes, such as $v_3(k)$, $v_3(k-1)$, $\rho_4(k)$, $\rho_4(k-1)$, $\rho_5(k)$, $\rho_5(k-1)$, $q_4(k)$, $q_4(k-1)$, and $q_5(k)$, $q_5(k-1)$. Through the analysis, the merit of the best first selection was 96.5%, which means that these 10 attributes had the highest impact on the entire traffic flow dynamics. As an example, Fig. 9.3 illustrates the efficiency of the regression analysis, comparing $q_5(k)$ and $\hat{q}_5(k)$ of the training set. It illustrates that most of the $q_5(k)$ are close to $\hat{q}_5(k)$.

In the following, two VisSim simulation scenarios demonstrate the efficiency of the coordination strategy. In the example, the coordination of r_1 and κ is used through the optimization strategy (9.47) with the constraints $r_{1,max} = 1500$ veh/h, $\Delta r_1 = 500$ veh/h, $\kappa_{max} = 50\%$, $q_{0,lim} = 3000$ veh/h, and $e_{lim} = 500$ veh/h. In the given highway example, $q_{5,max} = 2500$ veh/h is yielded through the empirical analysis.

The results of the simulation are found in Fig. 9.4. In the scenario, the average uncontrolled inflow q_0 is 2000 veh/h, while the average inflow demand on the controlled ramp $r_{1,dem}$ is 1000 veh/h. It means that the overall average inflow demand is 3000 veh/h, which saturates the system, whose maximum outflow is $q_{5,max} = 2500$ veh/h. Thus, if all of the inflow demands enter the highway, it results

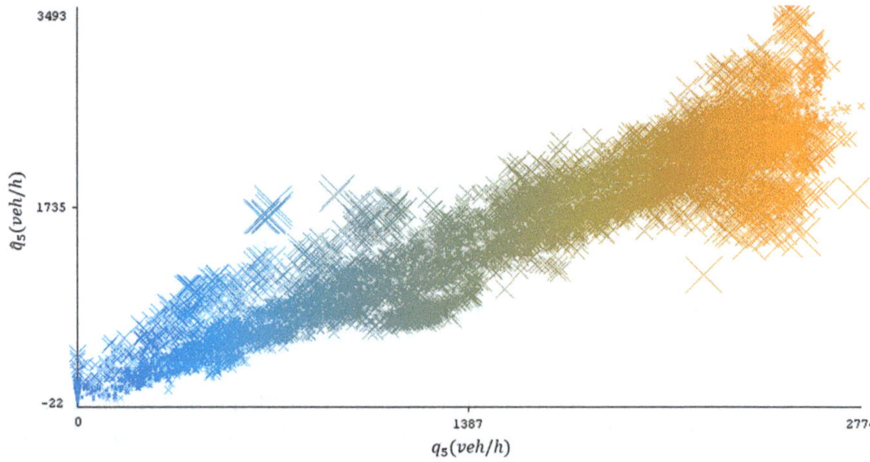

Fig. 9.3 Efficiency of the regression in case of q_5

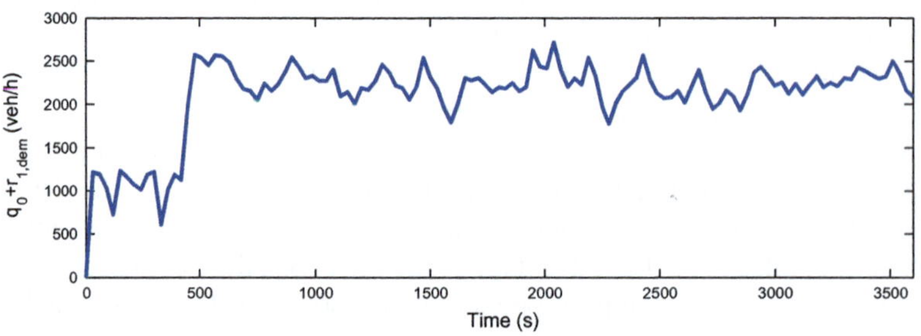

(a) Overall inflow value $q_0 + r_{1,dem}$ into the freeway

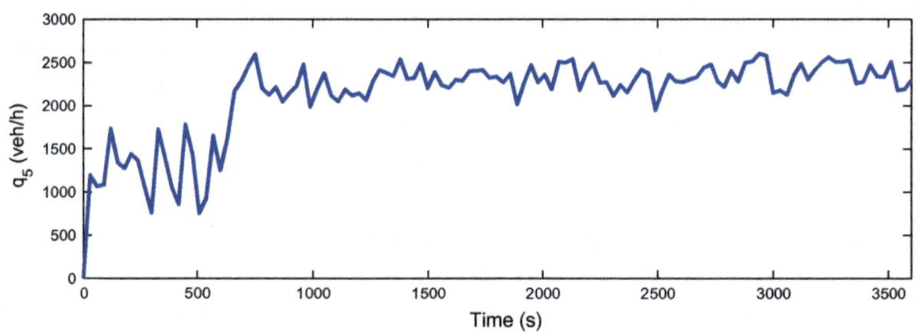

(b) Value of the outflow q_5

Fig. 9.4 Coordinated scenario—traffic flow

in a congestion. Therefore, it is necessary to actuate r_1 to avoid a traffic jam. At the beginning of the simulation, the inflow of the traffic is smaller, while in the next period of the simulation a rush-hour traffic develops, see the overall inflow $q_0 + r_1$ from the previous highway in Fig. 9.4a. Due to the control actuation of r_1, the over-

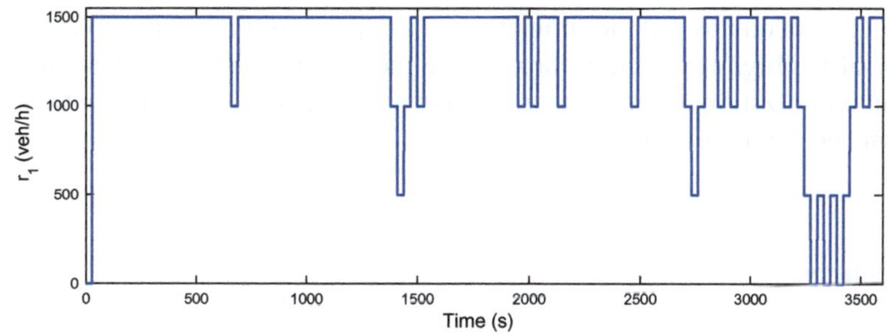

(a) Controlled inflow value r_1 into the highway

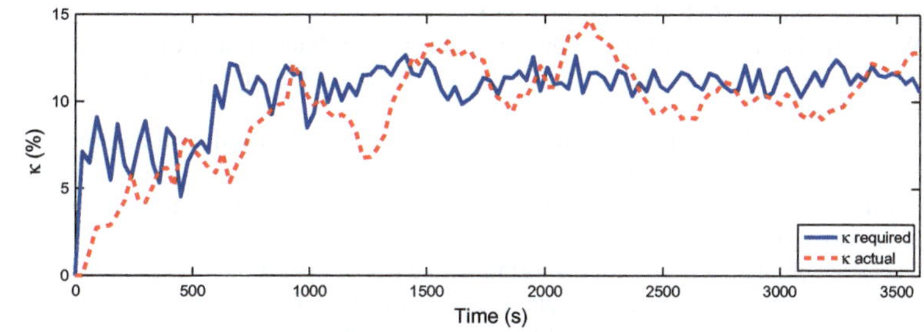

(b) The requested and the actual κ values

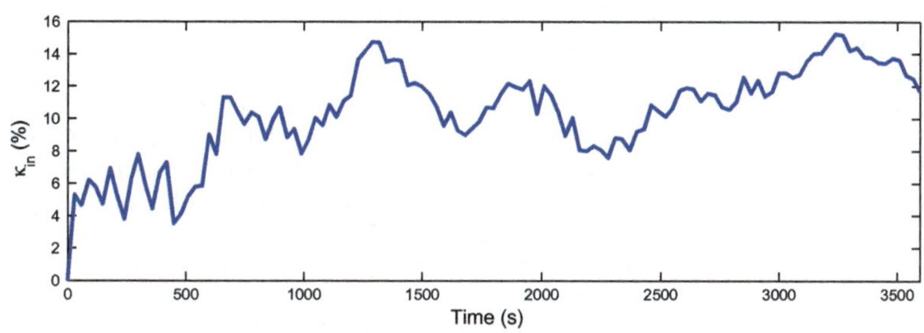

(c) Traffic flow density values

Fig. 9.5 Coordinated scenario—interventions

all inflow is limited, which results in the avoidance of congestion. The outflow of the highway section is shown in Fig. 9.4b. It can be seen that the reference tracking around $q_{5,max}$ can be guaranteed through the coordination.

The control actuation on the highway system is found in Fig. 9.5. The control input on the ramp r_1 is shown in Fig. 9.5a. Due to the rush traffic in the second part of the simulation, r_1 must be modified frequently to avoid congestion in the system. It can result in the limitation of the inflow vehicles on the controlled ramp. Moreover, the requested intervention of κ and the actual ratio of the eco-cruise controlled vehicles are shown in Fig. 9.5b. The limited error between the two signals illustrates the efficiency of the control strategy through κ_{in}, see Fig. 9.5c. Since there is a rush-hour traffic in the second part of the simulation, $\kappa < \kappa_{max}$ must be reduced to improve the outflow of the highway section through the increased speed of the vehicles. Thus, the joint optimization of the traffic ramp and automated vehicle interventions result in an enhanced coordination.

References

Amini S, Gerostathopoulos I, Prehofer C (2017) Big data analytics architecture for real-time traffic control. In: 5th IEEE international conference on models and technologies for intelligent transportation systems (MT-ITS), pp 710–715
Ashton W (1966) The theory of traffic flow. Spottiswoode, Ballantyne and Co. Ltd., London
Åström K, Hägglund T (2006) Advanced PID Control. ISA - The Instrumentation, Systems and Automation Society
Bose A, Ioannou P (2003) Analysis of traffic flow with mixed manual and semi-automated vehicles. IEEE Trans Intell Transp Syst 4(4):173–188
Coleman T, Li Y (1996) An interior, trust region approach for nonlinear minimization subject to bounds. SIAM J Optim 6:418–445
Daganzo C (1997) Fundamentals of transportation and traffic operations. Pergamon, Oxford
Dong H, Wu M, Ding X, Chu L, Jia L, Qin Y, Zhou X (2015) Traffic zone division based on big data from mobile phone base stations. Trans Res Part C 58:278–291
Frejo J, Camacho E (2012) Global versus local mpc algorithms in freeway traffic control with ramp metering and variable speed limits. IEEE Trans Intell Transp Syst 13(4):1556–1565
Gartner N, Wagner P (2008) Analysis of traffic flow characteristics on signalized arterials. Transp Res Rec 1883:94–100
Gill PE, Murray W, Wright M (1981) Practical optimization. Academic Press, London
Li L, amd Yanwei Wang XS, Lin Y, Li Z, Li Y (2015) Robust causal dependence mining in big data network and its application to traffic flow predictions. Transp Res Part C 58:292–307
Lv Y, Duan Y, Kang W, Li Z, Wang FY (2015) Traffic flow prediction with big data: a deep learning approach. IEEE Trans Intell Transp Syst 16(2):865–873
Messmer A, Papageorgiou M (1990) METANET - a macroscopic simulation program for motorway networks. Traffic Eng Control 31:466–470
Németh Gáspár P (2017) The relationship between the traffic flow and the look-ahead cruise control. IEEE Trans Intell Transp Syst 18(5):1154–1164
Németh B, Bede Z, Gáspár P (2015) Control design of traffic flow using look-ahead vehicles to increase energy efficiency. In: American control conference, Seattle, USA, pp 3447–3452
Papageorgiou M, Vigos G (2008) Relating time-occupancy measurements to space-occupancy and link vehicle-count. Trans Res Part C 16(1):1–17
Schakel W, Van Arem B, Netten B (2010) Effects of cooperative adaptive cruise control on traffic flow stability. In: 2010 13th international IEEE conference on intelligent transportation systems (ITSC), pp 759–764. https://doi.org/10.1109/ITSC.2010.5625133
Shibata R (1981) An optimal selection of regression variables. Biometrika 68:45–54
Thompson ML (1978) Selection of variables in multiple regression. Int Stat Soc B 46:1–21 and 129–146
Treiber M, Kesting A (2013) Traffic flow dynamics: data models and simulation. Springer, Heidelberg
van Arem B, van Driel CJG, Visser R (2006) The impact of cooperative adaptive cruise control on traffic-flow characteristics. IEEE Trans Intell Transp Syst 7(4):429–436
Wageningen F, Lint H, Vuik K, Hoogendoorn S (2015) Genealogy of the traffic flow models. Eur J Transp Logist 4:445–473

Wang Y, Witten IH (1999) Pace regression. (Working paper 99/12). University of Waikato, Department of Computer Science, Hamilton, New Zealand

Wang Y, Witten IH (2002) Modeling for optimal probability prediction. In: Proceedings of the nineteenth international conference in machine learning, Sydney, Australia, pp 650–657

Witten IH, Frank E (2005) Data mining practical machine learning tools and techniques. Morgan Kaufmann Publishers, Elsevier

Zeyu J, Shuiping Y, Mingduan Z, Yongqiang C, Yi L (2017) Model study for intelligent transportation system with big data. Procedia Comput Sci 107:418–426

Zhang C, Vahidi A (2011) Predictive cruise control with probabilistic constraints for eco driving. In: ASME 2011 dynamic systems and control conference and bath/ASME symposium on fluid power and motion control, vol 2, pp 233–238

Zhu M, Liu XY, Qiu M, Shen R, Shu W, Wu MY (2016) Traffic big data based path planning strategy in public vehicle systems. In: 2016 IEEE/ACM 24th international symposium on quality of service (IWQoS), pp 1–2

Chapter 10
Cruise Control Design in the Platoon System

Introduction and Motivation

In this chapter, the adaptive cruise control method will be extended to platoon systems. The term platoon is used to describe several vehicles operated under automatic control as a unit when they are traveling at the same speed with relatively small inter-vehicle spacings. Platoon operations may have advantages in terms of increasing highway capacity and decreasing fuel consumption.

The thoughts of platoon control were motivated by intelligent highway systems and road infrastructure, see the PATH program in California and the MOC-ITS program in Japan. The European programs were based on the existing road networks and infrastructure and focused mainly on commercial vehicles with their existing sensors and actuators, see Deutschle et al. (2010), Larburu and Sanchez (2010), Sitavancova and Hajek (2010). The main goal of the projects was to examine the operation of platoons on public motorways with full interaction with other vehicles. The safety and stability of each vehicle in the platoon is guaranteed by string stability, see e.g. Shaw and Hedrick (2007), Swaroop and Hedrick (1996), Alvarez and Horowitz (1999). In a Hungarian project an automated vehicle platoon of heavy vehicles was developed. The goal of the project was to analyze the control algorithms and synthesize the experimental results. Several questions which were related to junctions, coupling, driver behavior, technical failures, etc., had to be answered, see Rödönyi et al. (2012).

The purpose is to extend the string-stable platoon control with the predicted road conditions. Using the different performance specifications a control-oriented model of a platoon system is formulated. Uncertainties of the model, which are caused by neglected components, and unknown parameters are also modeled as unstructured dynamics. The controller of the platoon system is designed by using a robust $\mathscr{H}_\infty$ method, which guarantees both disturbance attenuation and robustness against uncertainties. The schematic structure of the controlled platoon system is shown in Fig. 10.1.

P. Gáspár and B. Németh, *Predictive Cruise Control for Road Vehicles Using Road and Traffic Information*, Advances in Industrial Control,
https://doi.org/10.1007/978-3-030-04116-8_10

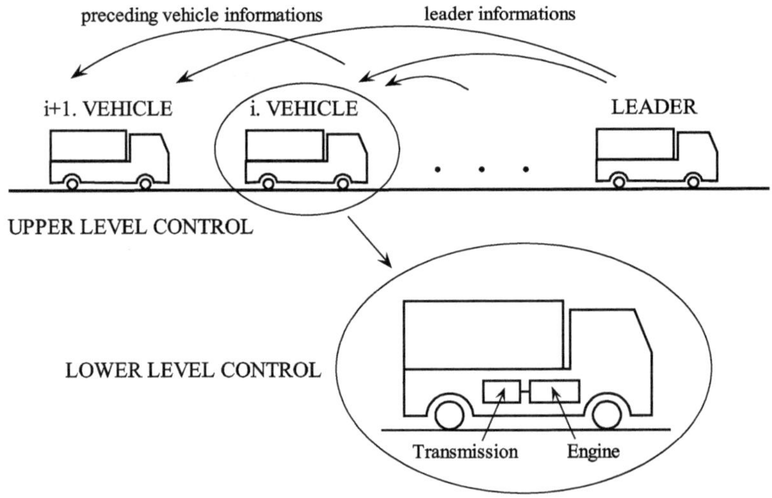

Fig. 10.1 Structure of a platoon system

A platoon control requires various measured signals, such as positions, velocities, and accelerations of the leader and the preceding vehicles. Thus, the platoon control requires that the position, velocity, and acceleration of a vehicle are transmitted to the following vehicle in order to ensure string stability. Besides these signals the road conditions are assumed to be available in the vehicles (road slope, speed limit). Since the safe and economical movement of the platoon is determined by the leader vehicle, it is crucial that the leader vehicle uses the predicted road conditions. From these signals the control system of each vehicle calculates its optimal longitudinal control force.

The chapter proposes a control design method for the platoon system. The controlled system incorporates the brake and the traction forces. The method takes safe travel into consideration by using the string stability theorem and the knowledge of the inclinations of the road along the route of the platoon and the compulsory speed limits. By choosing the velocity of the platoon fitting in with the inclinations of the road the number of unnecessary accelerations and brakes, i.e., the activation of the traction and brake systems, can be reduced.

10.1 Design of the Leader Velocity Based on an Optimization Method

In this section the method will be extended to vehicles in a platoon. The main idea behind the design is that each vehicle in the platoon is able to calculate its velocity independently from the other vehicles. Since traveling in a platoon requires the same velocity, the optimal velocity must be modified according to the other vehicles. In the platoon, the velocity of the leader vehicle determines the velocity of all the vehicles.

The goal is to determine the velocity at which the velocities of the members are as close as possible to their own optimal velocity.

In the first step the optimal prediction weights are set $\{\bar{Q}_j; \bar{\gamma}_{i,j}\}, i \in [1; n]$ where n is the number of division points. Then the modified reference velocities of all the vehicles λ_j, $j \in [1; m]$ where m is the number of the vehicles in the platoon are calculated. During the calculation the road inclinations and the speed limits are taken into consideration, however, the interaction between the vehicles are not considered. In the case of a platoon each vehicle has its own optimal reference velocity λ_j. Moreover, velocities of the vehicles are not independent of each other, because the velocity of the leader λ_1 influences the velocity of every member of platoon $\dot{\xi}_{0,j}$. The goal is to find an optimal reference velocity for the leader $\bar{\lambda}_1$.

Before the required velocity of the leader vehicle $\bar{\lambda}_1$ is designed a second—intermediate—step is needed. It is important to note that there is an interaction between the velocities of the vehicles in a platoon. If a preceding vehicle changes its velocity, the follower vehicles will modify their velocities and track the motion of the preceding vehicle within a short time. The members of the platoon are not independent of the leader, therefore it is necessary to formulate the relationship between the velocity of the jth platoon member $\dot{\xi}_{0,j}$ and the leader and the preceding vehicles. It is formulated with a transfer function with its input and output. The output

$$Y_j = \mathscr{F}\left[\mathscr{L}(\ddot{\xi}_{0,j})\ \mathscr{L}(\dot{\xi}_{0,j})\ \mathscr{L}(\xi_{0,j})\right]^T \tag{10.1}$$

contains the information, i.e., the position, velocity, and acceleration of the vehicle, which is sent to the follower vehicles. The input

$$U_j = \mathscr{G}\left[\mathscr{L}(\ddot{\xi}_{0,j-1})\ \mathscr{L}(\dot{\xi}_{0,j-1})\ \mathscr{L}(\xi_{0,j-1})\ \mathscr{L}(\dot{\xi}_0)\ \mathscr{L}(\xi_0)\right]^T \tag{10.2}$$

contains the position, velocity, acceleration of the preceding vehicle and the leader. The transfer function between U_j and Y_j is

$$G_{j,cl} = K_j G_j/(1 + K_j G_j) \tag{10.3}$$

with the controller of the jth vehicle K_j and its longitudinal dynamics G_j. Similarly, the effect of the leader and the preceding vehicles on the $(j+1)$th platoon member is formulated: $Y_{j+1} = G_{j+1,cl}U_{j+1}$, where $U_{j+1} = \begin{bmatrix} G_{j,cl} & \\ 0 & I \end{bmatrix} U_j$, which finally leads to

$$Y_{j+1} = G_{j+1,cl} \begin{bmatrix} G_{j,cl} & \\ 0 & I \end{bmatrix} U_j \tag{10.4}$$

Consequently, the velocity of the jth vehicle are determined by the next formula:

$$\mathscr{L}[\dot{\xi}_{0,j}] = \begin{bmatrix} 0 \\ 1 \\ 0 \end{bmatrix}^T Y_j = \begin{bmatrix} 0 \\ 1 \\ 0 \end{bmatrix}^T G_{j,cl} \prod_{k=2}^{j-1} \begin{bmatrix} G_{k,cl} & \\ 0 & I \end{bmatrix} G_{1,cl}\bar{\lambda}_1 = \hat{G}_j \bar{\lambda}_1 \tag{10.5}$$

The value of $\hat{G}_j$ is used for the computation of the optimal reference velocity of the platoon $\bar{\lambda}_1$.

In the third step the required reference velocity of the leader vehicle $\bar{\lambda}_1$ is designed. The aim of the design is that the generated velocities of all the vehicles $\dot{\xi}_{0,j}$ are as close to as their modified reference velocity λ_j as possible

$$\sum_{j=1}^{m} |\lambda_j - \dot{\xi}_{0,j}| \rightarrow Min. \tag{10.6}$$

Since the velocity of the jth vehicle is formulated as $\dot{\xi}_{0,j} = \hat{G}_j \bar{\lambda}_1$, the following optimization form is used:

$$\sum_{j=1}^{m} \left(\lambda_j - \hat{G}_j \bar{\lambda}_1\right)^2 = \sum_{j=1}^{m} \lambda_j^2 + \sum_{j=1}^{m} \left(\hat{G}_j \bar{\lambda}_1\right)^2 - 2\sum_{j=1}^{m} \left(\lambda_j \hat{G}_j \bar{\lambda}_1\right) \rightarrow 0 \tag{10.7}$$

where $\hat{G}_j = \prod_{k=1}^{j-1} \hat{G}_k$. It can be stated that in (10.7) the only unknown variable is $\bar{\lambda}_1$. The optimization leads to the following equation:

$$\bar{\lambda}_1 \sum_{j=1}^{m} \hat{G}_j^2 - \sum_{j=1}^{m} \left(\lambda_j \hat{G}_j\right) = 0. \tag{10.8}$$

The solution of the optimization problem can be achieved. The deduction of the optimization of $\bar{\lambda}_1$ is as follows:

$$\bar{\lambda}_1 = \frac{\sum_{j=1}^{m} (\lambda_j \prod_{k=1}^{j-1} \hat{G}_k)}{\sum_{j=1}^{m} (\prod_{k=1}^{j-1} \hat{G}_k)^2} \tag{10.9}$$

It means that the leader vehicle must track the required reference velocity $\bar{\lambda}_1$.

The calculation of the longitudinal force, which is illustrated in Fig. 10.2, is performed in three steps:

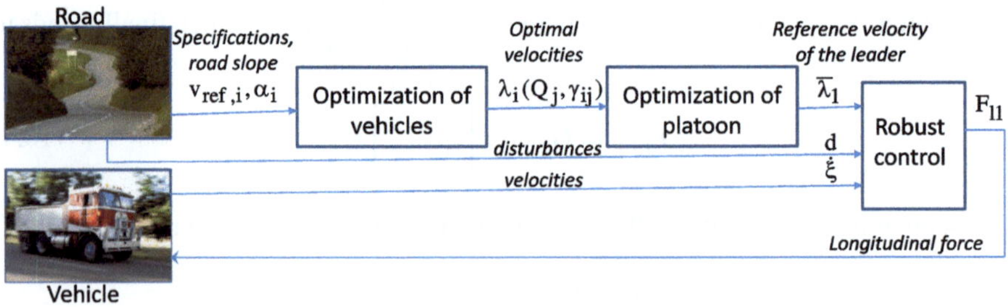

Fig. 10.2 Architecture of the control system

- The optimization of velocities of the vehicles is based on two criteria $|F_{l1}^2| \to Min$ and $|v_{ref0} - \dot{\xi}_0| \to Min$. The simplex algorithm is used.
- The optimization of the velocity of the leader vehicle $\sum_{j=1}^{m} |\lambda_j - \dot{\xi}_{0,j}|^2 \to Min$. A simple matrix manipulation is used.
- The required longitudinal force is based on the robust control method. The robust control is designed in an offline way.

10.2 Design of Vehicle Control in the Platoon

10.2.1 Design of Robust Control

This section focuses on the design of the control input, i.e., the design of the longitudinal forces of the vehicles in the platoon. In the previous section the required reference velocity of the leader vehicle was determined by taking the road inclinations into consideration. The velocity of the leader vehicle must track the required velocity. At the same time the other vehicles in the platoon must meet the string-stable requirement in order to guarantee the safe operation of the platoon. Consequently, two types of controllers must be designed: a velocity tracking controller for the leader vehicle and string-stable controllers for each vehicle in the platoon. All the controllers must provide disturbance attenuation and robustness against uncertainties.

The design of controllers are based on robust $\mathscr{H}_\infty$ methods. The control design of the leader vehicle is a tracking problem formulated in (2.32). The aim of tracking is to ensure that the system output follows a reference value of velocity with an acceptable error, which is the performance of the system. The explicit mathematical description of the optimization problem is $\bar{\lambda}_1 - \dot{\xi}_{0,1} \longrightarrow Min!$, where the parameter $\bar{\lambda}_1$ is the required reference velocity. The performance of the leader vehicle is as follows:

$$z_1 = [\bar{\lambda}_1 - \dot{\xi}_{0,1}] \tag{10.10}$$

The other vehicle in the platoon must guarantee string stability. This property of the system ensures that the inter-vehicular spacing errors of all the vehicles remain bounded uniformly in time, provided the initial spacing errors of all the vehicles are bounded. There are several strategies by which it is possible to ensure the string stability of the system. String stability is assumed to be ensure by tracking the position, velocity, and acceleration of the preceding vehicle, and tracking the position and velocity of the leader vehicle. The performance vector of the jth vehicle is

$$z_j = [\ddot{\varepsilon}_j;\ \dot{\varepsilon}_j;\ \varepsilon_j;\ \dot{\xi}_{0,j} - \dot{\xi}_{0,1};\ \xi_{0,j} - \xi_{0,1} + \sum_{i=1}^{j} L_i]^T, \tag{10.11}$$

where L_i is the distance between two vehicles and $\varepsilon_i = \xi_{0,j} - \xi_{0,j-1}$ is the error of tracking.

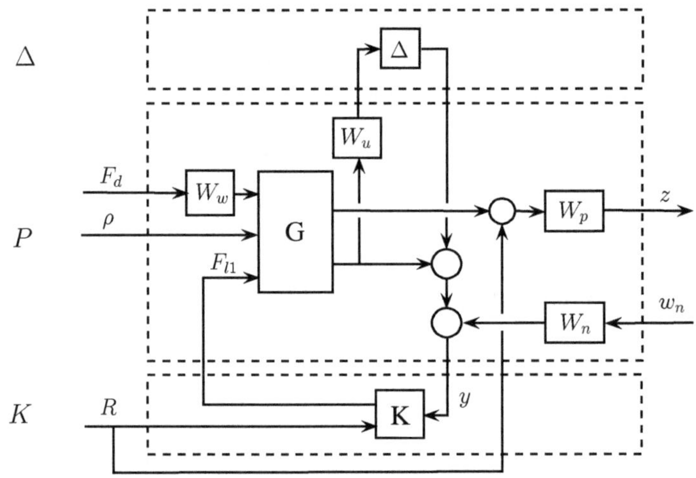

Fig. 10.3 Closed-loop interconnection structure of the system

The closed-loop interconnection structure, which includes the feedback structure of the model P and controller K, is shown in Fig. 10.3. The control design is based on a weighting strategy. The purpose of weighting function W_p is to define the performance specifications of the control system, i.e., the velocity of the vehicle must ensure the tracking of the reference signal with an acceptable error. The purpose of the weighting function W_n is to reflect the sensor noise, while W_w represents the effect of longitudinal disturbances. In the modeling an unstructured uncertainty is modeled by connecting an unknown but bounded perturbation block (Δ) to the plant. The magnitude of output multiplicative uncertainty is handled by a weighting function W_u.

The performance signal z differs in the two cases. In case of the leader $z = z_1$ according to (10.10) and in the other vehicles $z = z_j$ according to (10.11). For the leader vehicle the measured signals are the velocity and acceleration of the vehicle, while the follower vehicles also require the velocity, position and acceleration information about the preceding and leader vehicles. The reference signal R is the required reference velocity $\bar{\lambda}_1$ at the leader vehicle, while for the follower vehicles R vector contains the velocity and position of the leader and the preceding vehicles and the acceleration of the preceding one.

10.2.2 Stability Analysis of the Closed-Loop System

The computation of λ depends on velocity $\dot{\xi}_0$ and acceleration $\ddot{\xi}_0$ of the vehicle and exogenous signals. The velocity of the vehicle is considered at the computation of the reference velocity signal, which is a feedback in the closed-loop system. The result of reference signal computation influences the stability of the controlled system.

Since the relationship between the vehicle parameters and λ is nonlinear, a simplification is used in the analysis. In the following a transfer function N_j is introduced. It represents the relationship between momentary velocity $\dot{\xi}_{0,j}$ and its reference λ_j. This model with an autoregressive structure (ARX) is identified, see Ljung (1999):

$$y_j(q) = \frac{B_j(q)}{A_j(q)} u_j(q) + e(q) \tag{10.12}$$

where $B(q)/A(q)$ is the discrete-time model of the continuous system N_j, $y(q)$ and $u(q)$ are discrete signals, which are sampled from $\lambda_j(t)$ and $\dot{\xi}_{0,j}(t)$, while $e(q)$ is white noise. According to (10.5), it is possible to formulate a SIMO system with input $\bar{\lambda}_1$ and outputs $\dot{\xi}_{0,j}$, $j \in [1, n]$:

$$G_{pl} = \left[\hat{G}_1 \; \hat{G}_2 \; \ldots \; \hat{G}_n\right]^T \tag{10.13}$$

The illustration of the controlled systems is shown in Fig. 10.4. Then (10.9) is rearranged in a matrix form as

$$\bar{\lambda}_1 = \frac{\sum_{j=1}^{n}(\lambda_j \prod_{k=1}^{j-1} \hat{G}_k)}{\sum_{j=1}^{n}(\prod_{k=1}^{j-1} \hat{G}_k)^2} = G_\lambda \xi \tag{10.14}$$

where $G_\lambda = (G_\Pi G_\Pi^T)^{-1}(G_\Pi N)$ and $\xi = \left[\dot{\xi}_{0,1} \; \ldots \; \dot{\xi}_{0,n}\right]^T$ with $G_\Pi = \left[1 \; \hat{G}_1 \; \hat{G}_1 \cdot \hat{G}_2 \; \ldots \; \prod_{k=1}^{n-1} \hat{G}_k)\right]$, and $N = diag(N_1; N_2 \ldots N_n)$. G_{filter} is a low-pass filter, which smoothes $\bar{\lambda}_1$ reference signal. Note this function must be used because the necessary measurements for the computation of $\bar{\lambda}_1$ are sampled. G_{filter} is also constant.

Stability of the closed-loop system can be analyzed by the Small gain theorem, see Zhou et al. (1996)

$$\|G_{filter} G_\lambda\| \|G_{pl}\| < 1 \tag{10.15}$$

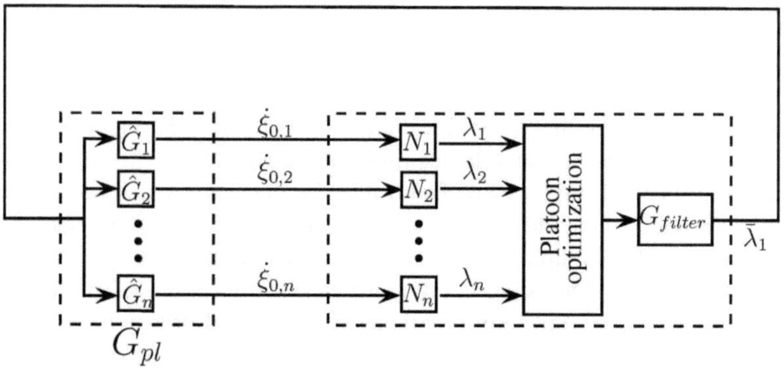

Fig. 10.4 Closed-loop system of platoon control

Equation (10.15) shows that the stability of the closed-loop system depends on the infinite norms of G_{pl}, G_{λ}, G_{filter}. Since these factors depend on dynamics and controllers of vehicles, infinite norm (10.15) depends on N, which is identified, see (10.12).

If (10.15) is not guaranteed by $\bar{\lambda}_1(k)$, then it is overwritten by $\hat{\lambda}_1(k-1)$, which is the predicted output signal for step k, which is computed in step $k-1$. It means that reference signal is constant. Note this analysis is relevant mainly in practice, when $\lambda > \dot{\xi}_0$, i.e., the vehicle is accelerating, e.g., before an uphill section or when a speed limit is lifted.

10.3 Simulation Results

In this section, a transportational route with real data is analyzed. The terrain characteristics and geographical information are those of the $A8$ German motorway between Ulm and Stuttgart in a 90 km long section. The motorway $A8$ runs across the Alps in South-Germany and connects Luxembourg with Austria, therefore it is one of the most important East–West transit route. The chosen road section, especially Swabian part contains uphills, in which the energy consumption of the platoon is critical. Publicly accessible up-to-date geographical/navigational databases and visualization programs, such as Google Earth and Google Maps, are used for the experiment. The platoon in the simulation contains four members, each of them are 3500 kg light commercial vehicle (LCV). The recommended maximal speed on the motorway is 130 km/h but the road section contains also speed limits (e.g., 50 or 80 km/h). The length of the predicted road horizon is 500 m, which is divided into 10 points.

Figure 10.5a shows the altitude of the road section along the way. In this simulation, two different controllers are compared. Controller 1 is a conventional platoon controller, which ignores the predicted road information, while Controller 2 is the proposed platoon controller, which considers the predicted road conditions such as the inclinations and speed limits. Figure 10.5c illustrates speed values of leader and fourth vehicles of the platoon using Controller 1. It tracks the predefined speed limits as accurately as possible and the tracking error is minimal. Controller 2 modifies the speed values considering speed limits and road inclinations simultaneously according to the optimal requirement, see Fig. 10.5d. For example, at 31 km of the road there is a speed limit, therefore the vehicles must be decelerated to 80 km/h. Controller 1 reduces the speed abruptly, while Controller 2 reduces the speed in a longer way exploiting the adhesion of the road. In another example at 34 km of the road there is a downhill section after an uphill. Controller 2 reduces the speed before the top of the section, because the vehicles are going to accelerate on downhill. The longitudinal control forces of leader and fourth vehicles are presented in Fig. 10.5e and f. The high-precision tracking of the predefined speeds in the conventional platoon control system often requires extremely high forces with abrupt changes in the signals.

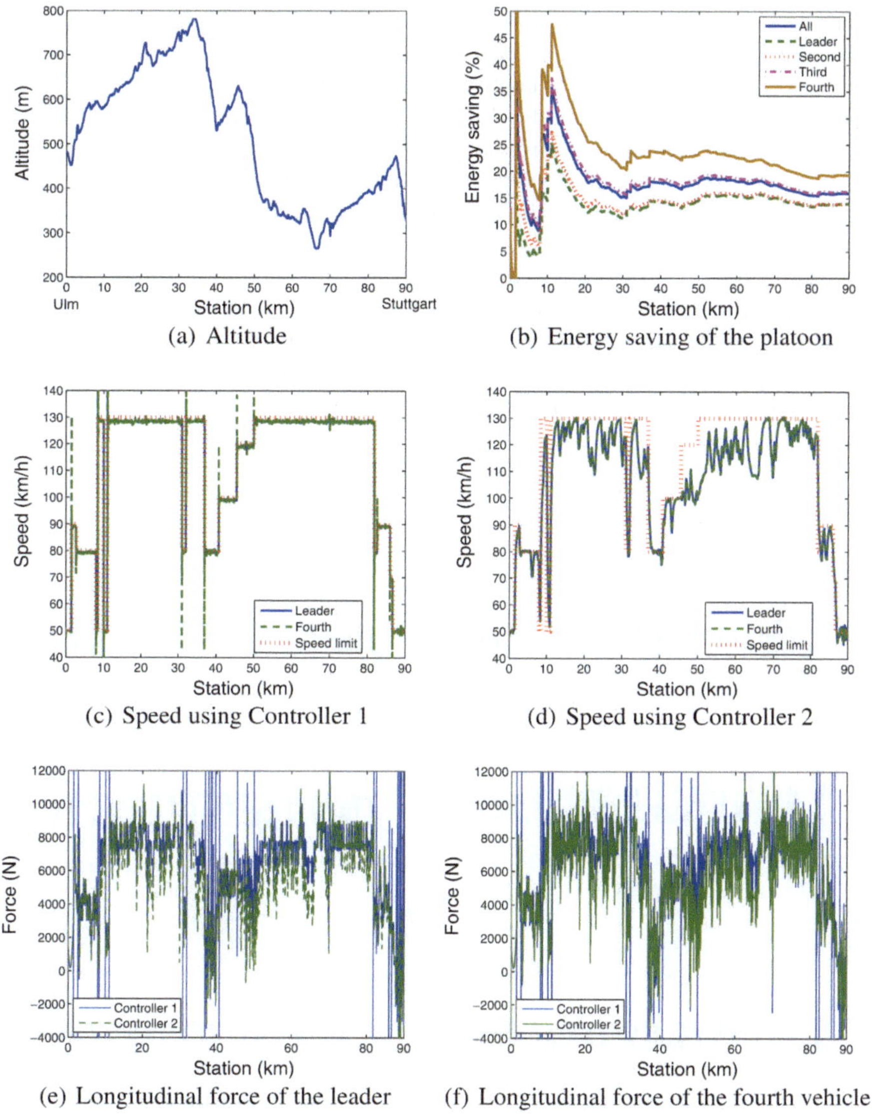

(a) Altitude

(b) Energy saving of the platoon

(c) Speed using Controller 1

(d) Speed using Controller 2

(e) Longitudinal force of the leader

(f) Longitudinal force of the fourth vehicle

Fig. 10.5 Real data motorway simulation

Since the proposed method calculates the longitudinal forces in advance based on the road inclinations and speed limits the truck is able to travel along the road with smaller actuation.

As a result of the predicted road conditions less energy is required during the journey in the proposed control method, see Fig. 10.5b. It is shown, that the saved energies are different at the platoon, in case of leader it is 14%, while at the fourth vehicle it is almost 20%. The overall energy saving of the platoon is 16%, which can be divided to brake actuation saving (88% of it) and powertrain actuation saving (12%). Fuel consumption can also be evaluated using the following simple equation: $V = E/(\eta L_h \rho_{fuel})$, where E is control energy, η is the efficiency of the powertrain system, L_h is heating value and ρ_{fuel} is density of gasoline. This approximation results 5.9 l reduction in fuel consumption for the entire platoon in the 90 km length section. Since in the method the velocity of the vehicle may be below the permitted maximum for the given section and accelerations/decelerations are carried out more slowly and gradually than in the conventional method duration of the journey is expected to be longer. However, the difference in the duration is only 2 minutes. Therefore the increase of travel time is 3.9%.

Stability of the platoon system is analyzed by using Eq. (10.15). The analysis requires the identification of the systems N_j, see Eq. (10.12) and Fig. 10.4.

The transfer functions N_j are considered in the form

$$N_j = \frac{B_j(q)}{A_j(q)} = \frac{1 + b_1 q^{-1} + b_2 q^{-2}}{1 + a_1 q^{-1} + a_2 q^{-2}} \tag{10.16}$$

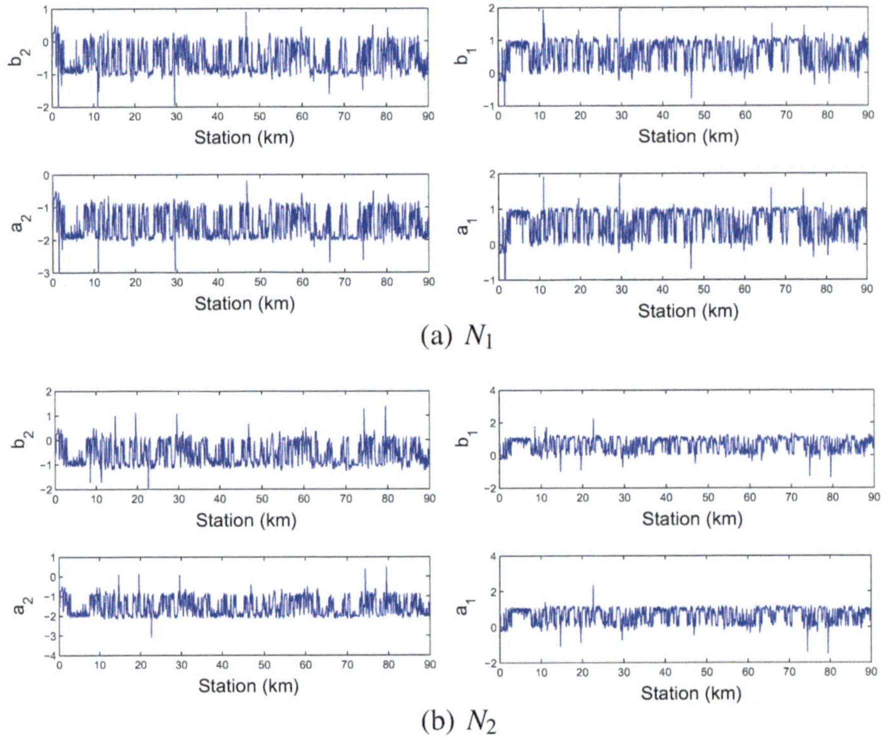

Fig. 10.6 Parameters of the N_j transfer functions

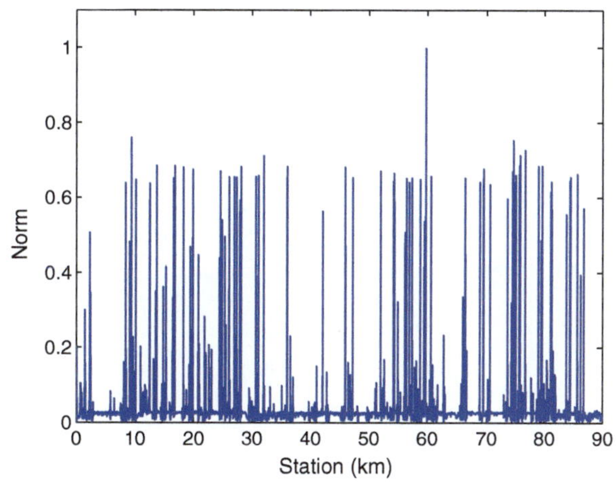

Fig. 10.7 Stability norm of the platoon

where a_1, a_2, b_1, b_2 are the identified parameters. Since $\dot{\xi}_{0,j}$ and λ_j varies during the cruising of the vehicles, the parameters must be identified continuously. As an example the parameter values of the transfer functions N_1 and N_2 are illustrated in Fig. 10.6.

Using the transfer functions N_j the robust stability of the platoon system can be analyzed through the norm criteria, see Eq. (10.15). In the example Fig. 10.7 shows the calculated norm of the closed platoon system. The values of the norm are under the critical level 1, which guarantees stability of the system by exploiting the Small gain theorem. This example has shown that the proposed method is able to save energy through the predictive cruise control.

References

Alvarez L, Horowitz R (1999) Safe platooning in automated highway systems part I: safety regions design; part II: velocity tracking controller. Veh Syst Dyn 32(1):23–84

Deutschle S, Kessler G, Hakenberg M, Abel D (2010) The konvoi project: development and investigation of truck plaoons on highways. FISITA Congress, Budapest

Larburu M, Sanchez J (2010) Safe road trains for environment: human factor aspects in dual mode transport systems. In: 17th world congress on intelligent transport systems, Busan, Korea

Ljung L (1999) System identification: theory for the user. Prentice Hall

Rödönyi G, Gáspár P, Bokor J, Aradi S, Hankovszki Z, R Kovacs LP (2012) Guaranteed peaks of spacing errors in an experimental vehicle string. In: 17th IFAC symposium on robust control design, Aalbord, Denmark

Shaw E, Hedrick J (2007) String stability analysis for heterogeneous vehicle strings. In: American control conference, pp 3118 – 3125

Sitavancova Z, Hajek M (2010) Promote-chauffeur: intelligent transport systems. EU Mobility and Transport

Swaroop D, Hedrick J (1996) String stability of interconnected systems. IEEE Trans Autom Control 41:349–357

Zhou K, Doyle J, Glover K (1996) Robust and optimal control. Prentice Hall

Chapter 11
Simulation and Validation of Predictive Cruise Control

11.1 Architecture of the Vehicle Simulator

This chapter deals with the development, testing, and implementation process of the look-ahead cruise control method. During the development of the look-ahead control system, building of a real-time vehicle simulator is based on a high-accuracy vehicle simulation software, see Mihály et al. (2017). Hence, the embedded control system could be tested for different scenarios without the use of a prototype vehicle. Here, both software-in-the-loop (SIL) and hardware-in-the-loop (HIL) simulations have been performed in order to verify the control algorithm. These methods have being widely used by automotive companies, see Bringmann and Krämer (2008), Vandi et al. (2014). For instance, the HIL platform served as the basis for the improvement of driver assistant and autonomous systems in Gietelink et al. (2006), Deng et al. (2006). Here, the developed real-time HIL vehicle simulator has been used for multiple purposes. Apart from the real-time testing and tuning of the look-ahead algorithm, the control systems developed on different hardware and software units could be coordinated on this platform in order to function properly on a test truck at the end of the development process.

During the development phase of the look-ahead cruise controller system, a vehicle simulator was built in TruckSim in order to conduct real-time tests. These real-data simulator tests serve as a substitution for otherwise expensive and time-consuming real test drives with the implemented system, hence they increase the efficiency of the development process.

The architecture of the complete vehicle simulator containing elements of the HIL environment is depicted in Fig. 11.1. Here, the Simulator block represents the driver environment containing hardware components for representing the vehicle: the driver's seat, a force feedback steering wheel, the brake, the clutch and the accelerator pedals, gearbox, and control buttons. The vehicle dynamics and graphical visualization are generated by the real-time version of TruckSim, with the driver view projected in front of the driver. Hence, interventions of the test drivers are based on

P. Gáspár and B. Németh, *Predictive Cruise Control for Road Vehicles Using Road and Traffic Information*, Advances in Industrial Control,
https://doi.org/10.1007/978-3-030-04116-8_11

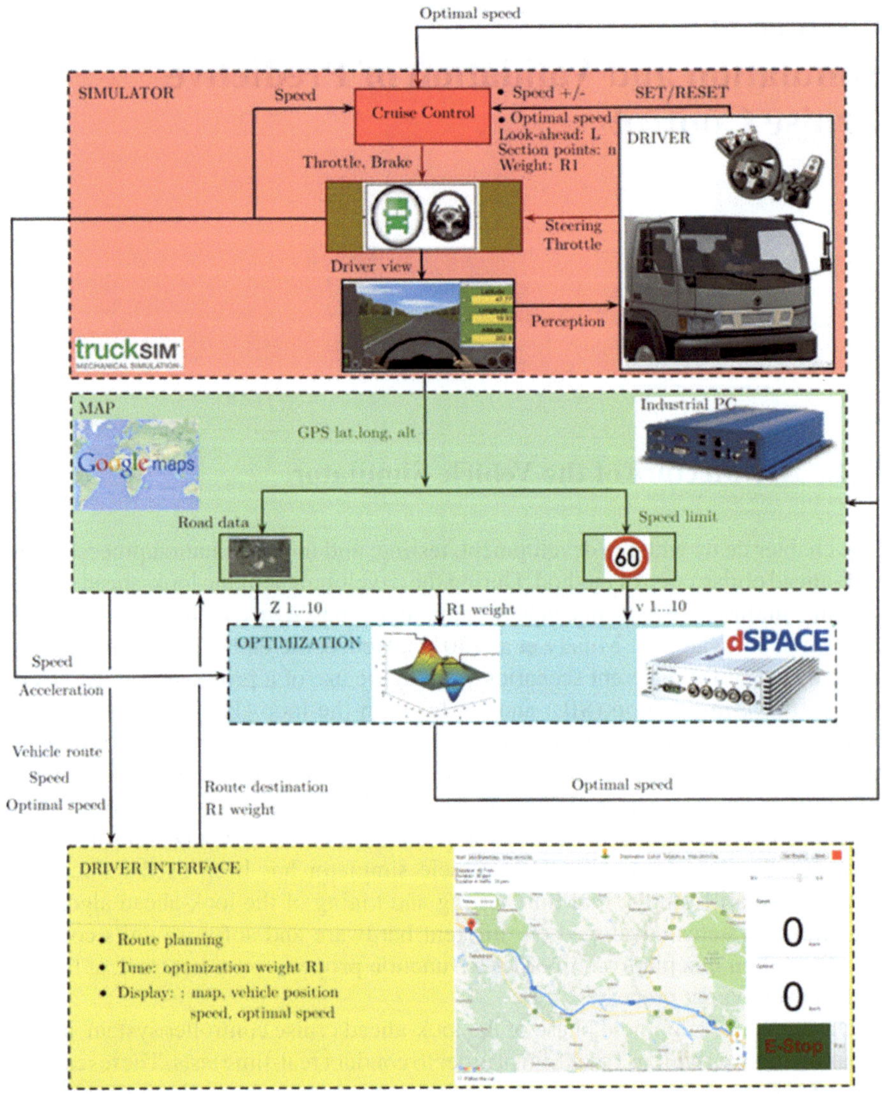

Fig. 11.1 Simulator flow chart

the visual and audio generation of the real-time vehicle simulator. Note that the simulated GPS signals representing the actual position of the vehicle are sent to the Map block, while the measured vehicle dynamic signals (velocity, acceleration) are transmitted to the Optimization block, in which the look-ahead control algorithm is running.

The role of the Map block is to provide forward road information for the look-ahead control algorithm, knowing the actual GPS position of the simulated vehicle.

Hence, for the predefined look-ahead distance terrain characteristic and speed limit data are downloaded from Google Maps. These data containing road information (slope angles, speed limits) for the look-ahead section points are then sent to the Optimization block.

Based on the measured simulated vehicle signals (acceleration, velocity) and the look-ahead road data provided by the Map block, the Optimization block solves the numerical equations and calculates the energy-optimal velocity for the vehicle. Thus, if the driver activates the look-ahead controller while driving the simulator, the energy-optimal velocity is given as reference velocity for the PID speed controller in the Simulation block.

The Driver Interface block serves to realize a two-way communication between the test driver and the control system. Hence, using a tablet computer the driver selects the destination and sets the value of energy optimization weight R_1. The interface shows the driver important information about the journey: current vehicle position on map, estimated traveling time, current speed and the calculated optimal speed.

The simulation environment is based on the real-time version of TruckSim, a high-fidelity vehicle simulator containing validated models of heavy-duty vehicles. Hence, the dynamics of the vehicle is calculated by using complex models of the engine, powertrain, braking system, etc. The MATLAB/Simulink model file includes an S-function block, which receives a file package from TruckSim containing data related to parameters of the selected vehicle and road as well as the traffic environment. During the real-time simulation, this S-function block is called and an animation of the vehicle motion is also generated by TruckSim. The simulation environment developed in the Simulator block is shown in detail in Fig. 11.2.

Note that the road information with the terrain characteristics, speed limits and curvature of the road is predefined in TruckSim based on the Google Maps database.

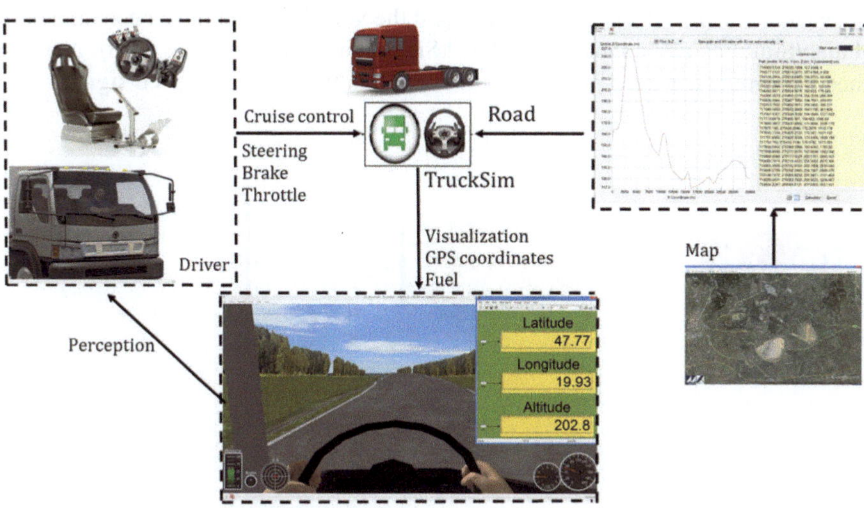

Fig. 11.2 Real-time simulation with TruckSim

The driver sitting in the simulator seat can choose between different driving modes and accordingly generates different inputs based on its audiovisual perception.

In order to realize cruise control functionality, a simplified vehicle model with appropriate linearization must be created for the longitudinal control of the vehicle. Here, a PI speed controller has been designed and tuned with a feedback linearization technique, thus the strongly nonlinear components of the vehicle dynamics (air resistance, slope resistance, rolling resistance) are calculated using the measured characteristics and added to the value driven by the controller. The advantage of this method is that the nonlinear system can be controlled by a simple linear PI regulator by separating nonlinear values that can be approximated. The speed controller calculates a longitudinal force that is realized by the engine and braking system of the vehicle, thus the engine butterfly valve position and the pressure of the vehicle master cylinder are adjusted.

For the longitudinal control of the simulated vehicle a simplified vehicle model is introduced. Since the output of the PI speed controller is a positive or negative longitudinal force, the low-level controllers must address the corresponding physical actuators (see Fig. 2.9). Thus, for setting the desired acceleration the throttle angle of the engine must be adjusted, while for braking the brake pressure must be set. Hence, by linearizing the vehicle model around the operation points, the actuator inputs can be defined accurately to meet the desired acceleration given by the reference longitudinal force of the speed controller.

During the PI control design a feedback linearization method has been applied. Thus, the highly nonlinear components of the vehicle dynamics (aerodynamic drag, road slope resistance, rolling resistance) are calculated by using measurement data. These nonlinear resistance forces are added to those provided by the speed controller. The method has the advantage that by separating the calculated nonlinearities a simple linear PID controller can be realized for the highly nonlinear system. The feedback linearized PI speed controller and the vehicle model have been created in Matlab/Simulink and connected to TruckSim environment (see Fig. 11.3). A predefined hilly route has been provided with a jump in the reference speed signal.

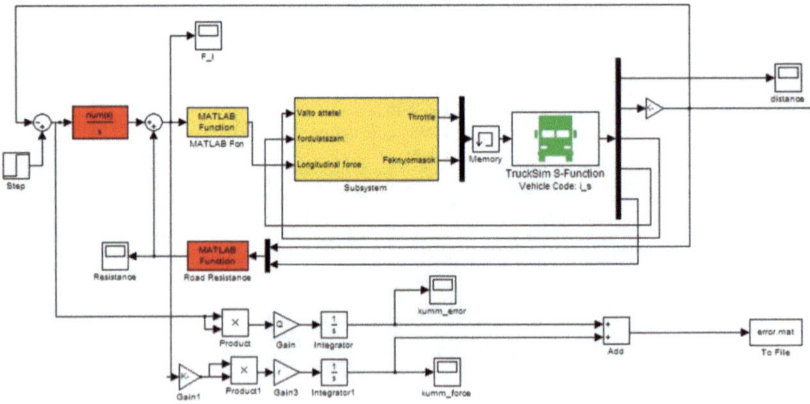

Fig. 11.3 Speed controller design

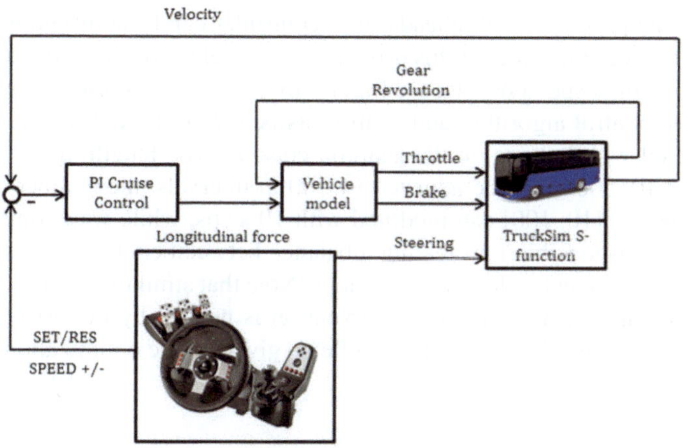

(a) Conventional cruise control mode

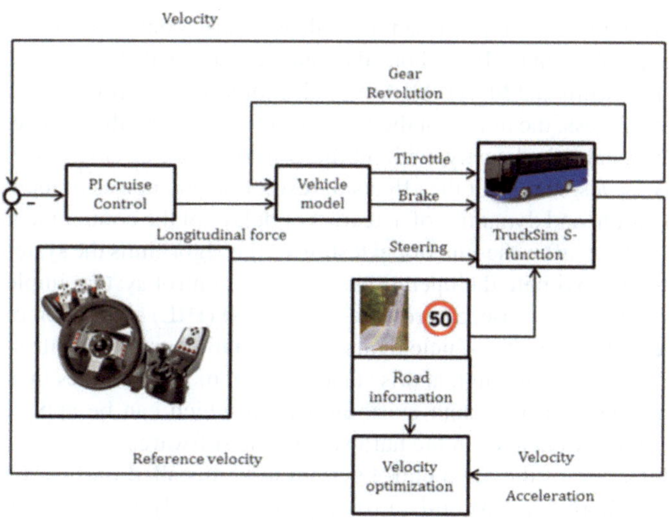

(b) Look-ahead cruise control mode

Fig. 11.4 Simulator driving modes

Two driving modes have been implemented for the simulator, which can be selected by pressing the appropriate button on the steering wheel. First, the designed conventional cruise control can be activated by pressing a steering wheel button, hence the vehicle follows a reference velocity set by the driver, see Fig. 11.4a. Moreover, this reference speed can be altered by 1 km/h intervals by pushing the left/right steering wheel buttons. Note that in this driving mode only steering manipulation is required by the driver.

Second, the proposed look-ahead cruise controller can be activated with another button on the steering wheel. If this driving mode is selected, the heavy-duty vehicle follows an optimal speed profile, see Fig. 11.4b. Moreover, in order to tune the proposed cruise control algorithm and analyze its behavior, three different parameters can be altered with buttons on the gearbox console. The length of the look-ahead horizon $L \in [0, 5000]$ can be adjusted by 500 m intervals, the number of segmentation points $n \in [10, 100]$ can modified with 10 steps, while weighting parameter $R_1 \in [0, 1]$ is responsible for creating a balance between energy-optimal operation and traveling time can be altered by 0.1 steps. Note that similarly to the conventional cruise control mode, only steering intervention is needed by the driver, while the reference velocity for the PI speed controller is given by the optimization algorithm.

11.2 Implementation of the Cruise Control on a Real Truck

The implementation of the proposed look-ahead cruise control system is evaluated in an incremental manner. First, both the vehicle itself and the control system are realized in TruckSim and MATLAB/Simulink simulation environment. At this stage of the design process, the tuning of the look-ahead control and the vehicle speed controller can be established easily, without the consideration of environmental effects or feasibility issues. Note that the effects of resolution and noises of the vehicle sensors, quantization, and dynamics of actuators or delays of the communication system were studied. Next, after the functional testing of the algorithms the system is further developed to approximate the operation of the real control system implemented on the vehicle. For this purpose, a hardware-in-the-loop (HIL) simulation environment has been created. Here, the vehicle dynamics simulation and the control system are physically separated. Although at this stage the system still operates in a simulation environment, the control system exists in a form which can be applied on a real vehicle without any changes in the hardware or the software.

The architecture of the real control system to be installed on-board is shown in Fig. 11.5. The MicroAutoBox on-board computer is directly connected to the vehicle via the CAN communication system and is able to control the speed of the vehicle through a custom-developed cruise control system. Note that the proposed look-ahead algorithm runs on MicroAutoBox as well as the cruise control system. This computer is also connected to a second industrial on-board computer via RS-232 interface, which is used to store or query online maps and other databases. It is implemented via a wireless communication unit through a GSM mobile Internet connection. The second on-board computer is responsible for receiving data from the GNSS receiver and transmitting it to MicroAutoBox in order to evaluate the look-ahead algorithm. This unit is responsible for communicating with a tablet computer via WiFi network, which performs the display and also serves as a user interface.

In the HIL environment, vehicle dynamics is simulated in the TruckSim environment, which also generates virtual GPS signals (longitude, latitude, altitude) for the industrial computer. Note that the protocol of the GNSS receiver used in

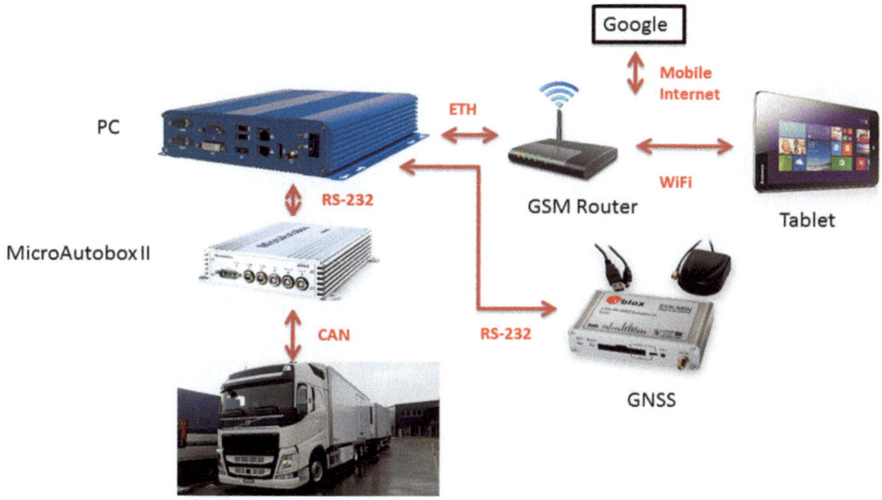

Fig. 11.5 Architecture of the control system

the real vehicle is emulated on the desktop computer simulating vehicle dynamics. Here, the desktop computer transmits the simulated velocity and acceleration data to MicroAutoBox via the CAN communication using the Vector CANCase card built in the desktop computer and the CAN interface of MicroAutoBox. The industrial on-board computer is connected via Ethernet and a WiFi Router to the internet databases (Google Maps) and sends the predicted road information to the MicroAutoBox. The actual and calculated optimal velocities along with the road map and current position of the vehicle are sent to the tablet application via WiFi communication. The calculated optimal velocity of the look-ahead algorithm implemented in MicroAutoBox is sent to TruckSim via CAN communication. Hence, in the HIL environment the structure of the control system conforms to the structure applicable to the real vehicle, with the exception of the physical absence of sensors for the vehicle motion state and the physical interfaces. The properties of the developed HIL system ensure that the hardware and software components running in the HIL simulation can operate unchanged on the actual vehicle on-board. Each part of the presented HIL control architecture has different tasks, see Fig. 11.6.

The MATLAB/Simulink environment is available for developing the MicroAutoBox computer program. From the generated schema, the development environment generates the program code that the device can run, downloaded to the device immediately after power-up. The basic task of the implemented software is to run the optimization procedure and to determine the optimal reference speed for the given point of the route. There are three main parts of the Matlab/Simulink model: receiving, processing, and transmitting. The receiving blocks of the model are responsible for receiving messages on the CAN network. These messages come from the simulator and include the current position, speed and acceleration of the vehicle, and

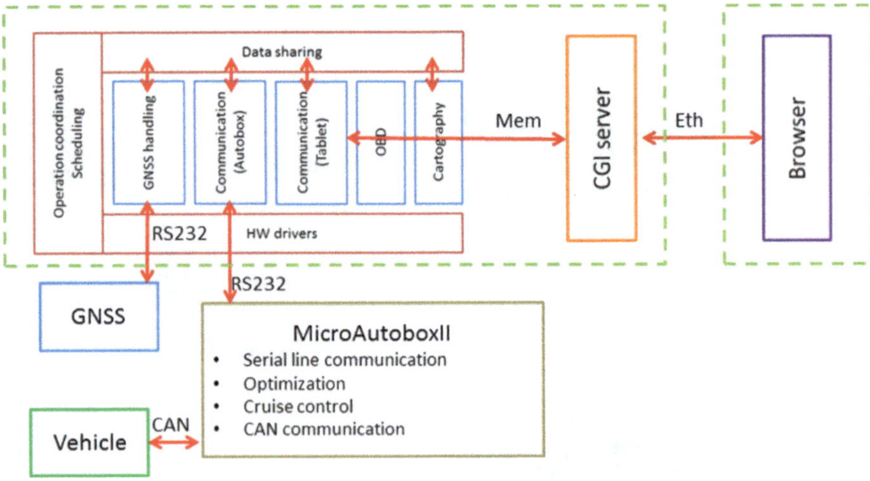

Fig. 11.6 Tasks of the different control system components

three tuning parameters (L, R_1, n) of the optimization algorithm. The processing block is responsible for determining the optimal reference speed. Note that altitude and speed limit data required to determine the optimal reference speed are available from the industrial on-board computer.

The calculated reference speed is sent via a weighting block in a CAN message to the simulator, which applies it to adjust the vehicle speed. The transmitting blocks transmit the current speed of the vehicle and the calculated reference speed over the serial line to the industrial on-board computer. It is necessary to define a planned route to operate the look-ahead cruise control system. As soon as the driver selects the destination using the tablet interface, the route is retrieved through Google Maps API. The encoded route points are then loaded into the central framework software.

The interface for the driver is a web-based application that can be viewed in a browser. Through this interface, the driver can select the route and start or stop the system. During the journey, current position and vehicle speed along with the optimal speed recommended by the look-ahead control system appear on the screen of the tablet. In addition to the most necessary information, the current local weather is also displayed on the map.

The main goal of the development was to create a HIL simulation architecture completely representing the real environment so that the developed system could be removed and inserted in the real vehicle. The hardware elements of the look-ahead speed control algorithm have been installed in the cockpit of the test truck. The on-board system includes the MicroAutbox computer, the industrial on-board computer, the GNSS system and the tablet computer.

The connection to the truck on-board management system is realized via the CAN network. Through this, the measurement data of the vehicle can be extracted and the traction power demand for the ECU controlling the engine and a deceleration command for the brake system can be obtained. Additionally, since intervention in the truck's cruise control system is not possible, the software implemented on MicroAutoBox computer is complemented with a custom-designed speed control algorithm.

In order to evaluate the results of the experimental tests and validate the effectiveness of the look-ahead method, it is necessary to measure the fuel consumption of the test vehicle. The measurement of the consumption is based on a message received via a further CAN channel directly from the CAN network of the vehicle.

The testing of the look-ahead control system has been carried out. The test vehicle used in the experiment is a three-axle truck. The total mass of the truck with the cargo was set at 16 tons, its power is 324 kw, while its geometric data are 9.3 m length, 2.5 m width and 3.7 m height.

In the case of the real truck it is necessary to guarantee the tracking of the computed energy-optimal speed profile through the real physical actuators of the vehicle. During the experimental validation the driveline management and the braking control systems of the truck were used. It means that it is necessary to design a cruise control whose outputs are the deceleration command for the braking control at speed reduction and torque command for the driveline management. Using the deceleration command the braking control generates the necessary air pressure on the brake chambers of the wheels. Moreover, the driveline management system controls the engine throttle and the transmission of the truck to guarantee the required torque command.

In the validation process of the predictive cruise control, a PI controller was designed to compute the deceleration and torque commands, see e.g., Rajamani (2005). The control strategy has two layers, such as the upper controller and the lower controller. The purpose of the upper controller is to compute the desired acceleration of the vehicle from the reference speed signal. Furthermore, the goal of the lower level control is to generate the torque command input for the driveline management from the desired acceleration signal.

The upper controller is found in a PI structure, such as

$$C(s) = k_p + \frac{k_i}{s}, \tag{11.1}$$

where k_p is the proportional gain and k_i is the integral gain. The requested tracking functionality is considered in a simplified form, such as

$$P(s) = \frac{1}{s(\tau s + 1)}, \tag{11.2}$$

where τ is the time constant related to the longitudinal vehicle dynamics. Generally, this parameter is around $\tau = 0.5$ s. The result of the controller is the desired acceleration $\ddot{\xi}_{des}$, while its input is the speed tracking error $\lambda - \dot{\xi}$. If $\ddot{\xi}_{des} \leq 0$ then $\ddot{\xi}_{des}$ is the input of the braking control. However, if $\ddot{\xi}_{des} > 0$ then it must be transformed through the low level controller to the input signal of the driveline.

In the case of the low-level controller the torque command for the driveline system must be calculated. Since the dynamics of the driveline incorporates several nonlinearities and uncertain parameters, the determination of the driveline torque is based on a simplified model. The longitudinal traction force of the vehicle is formed as

$$F_l = m\ddot{\xi}_{des} + F_r + F_{aer} + G_x, \tag{11.3}$$

as presented in Chap. 2, see (2.1) and (2.3). Using the relations (2.4)–(2.6), the traction force is expressed as

$$F_l = m\ddot{\xi}_{des} + F_z f_0(1 + \kappa\dot{\xi}^2) + 0.5C_w\rho A_0\dot{\xi}_{rel}^2 + mgsin\alpha. \tag{11.4}$$

Then, the desired longitudinal force F_l must be transformed into the torque of the driveline output shaft. The dynamics of the wheels is expressed as

$$\begin{aligned} I_w\dot{\omega}_w &= \frac{\ddot{\xi}_{des}}{r_{eff}} I_w = T_w - r_{eff} F_l = \\ &= T_w - r_{eff}\left(m\ddot{\xi}_{des} + F_z f_0(1 + \kappa\dot{\xi}^2) + 0.5C_w\rho A_0\dot{\xi}_{rel}^2 + mgsin\alpha\right), \end{aligned} \tag{11.5}$$

where I_w is the wheel inertia, $\dot{\omega}_w$ is the wheel acceleration, r_{eff} is the effective radius and T_w is the torque on the wheel. The relationship between the wheel and the output shaft of the driveline is determined by the final drive, whose ratio is R. Thus, the angular speed of the output shaft of the driveline is

$$\omega_{dr} = \frac{\omega_w}{R} \tag{11.6}$$

and the output torque of the driveline $T_{dr} = RT_w$ is

$$T_{dr} = \frac{R\ddot{\xi}_{des}}{r_{eff}} I_w + Rr_{eff}\left(m\ddot{\xi}_{des} + F_z f_0(1 + \kappa\dot{\xi}^2) + 0.5C_w\rho A_0\dot{\xi}_{rel}^2 + mgsin\alpha\right). \tag{11.7}$$

During the tuning of the speed controller, various parameters of the test truck were displayed on an on-board laptop computer. By this means, the speed profile of the truck, the motor torque demand and its realization also became visible. Following the tuning of the speed controller, the route selection and the timetable for conducting the road tests were determined.

11.2.1 Test Results

For testing the presented look-ahead cruise control algorithm, a 50-km-long section of the M1 motorway route between Tatabánya and Budapest has been chosen. This

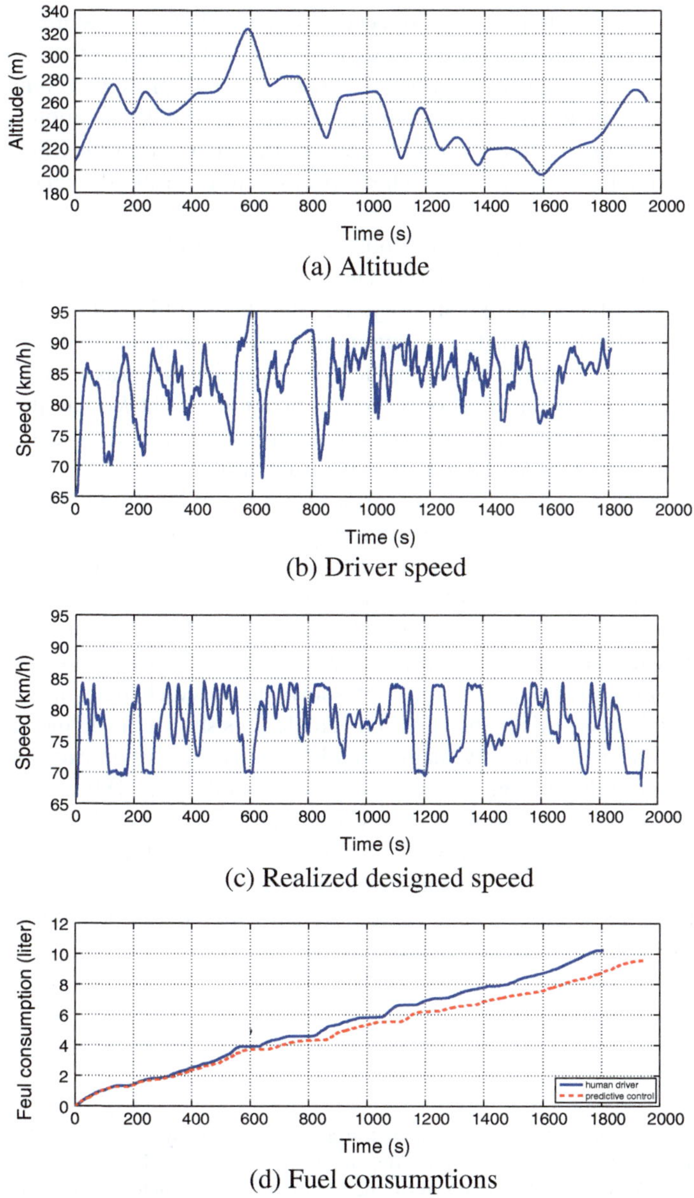

(a) Altitude

(b) Driver speed

(c) Realized designed speed

(d) Fuel consumptions

Fig. 11.7 A test result on the Tatabánya-Budapest road

road has been selected owing to its frequent use in freight transport and its diverse terrain characteristics with several uphill and downhill slopes.

Since additional speed limitation for the truck at the time of the measurement is not considered, the reference speed on the entire route was 80 km/h for the truck. Hence, it is important to note that the experienced differences in fuel consumption resulted only from the exploitation of the terrain characteristics. The measurements took place late at night, so the traffic did not affect the operation of the presented control system. During the measurements, the following data were saved: GPS position of the truck (longitude, latitude, altitude), current speed of the truck based on GNSS receiver, current speed of the truck based on wheel speeds, reference speed, fuel consumption of the truck.

The results of the implemented look-ahead control algorithm were compared those of an experienced truck driver. The driver traveled in his own driving style without the use of any cruise control system, thus the throttle and brake pedal were operated manually. Based on the measurements carried out in the preliminary vehicle simulations in TruckSim environment, the look-ahead distance was set at $L = 2000$ m with $n = 20$ section points, thus only the optimization weight R_1 was altered during the algorithm testing. Several test runs had been performed with different R_1 weights, and the results were compared to that of the truck driver.

As an illustration Fig. 11.7 shows a test result, i.e., the altitude, the speed of the driver, the realized designed speed and the fuel consumption. In the analysis the speed designed by the predictive cruise control and implemented by the truck is compared to the speed of the driver. Meanwhile in both cases the fuel consumptions are measured. The optimal weight is selected $R_1 = 0.85$. According to Fig. 11.7c the designed and implemented speed changes around 80 km/h, which is suitable for the speed limit of the truck. As Fig. 11.7b shows the speed changes in a wider range when the truck driver drives the common vehicle. It is shown the defined speed limit of 80 km/h was not accurately maintained by the driver, it often exceeded or gone below by the test driver. The entire journey was completed in 1807 s while the fuel consumption of the truck was 10.23 l. These metrics serve as a benchmark for the design method that was tested with multiple parameter settings. The journey was completed in 1944 s, while the fuel consumption of the truck was 9.37 l. Thus, by utilizing the proposed look-ahead control system with the aformentioned parameter settings, fuel consumption was reduced by 8.4%.

References

Bringmann E, Krämer A (2008) Model-based testing of automotive systems. In: IEEE international conference on software testing, verification, and validation, pp 485–493

Deng W, Lee Y, Zhao A (2006) Hardware-in-the-loop simulation for autonomous driving. In: 2008 IECON 34th annual conference of IEEE industrial electronics, pp 1742–1747

Gietelink O, Ploeg J, Schutter BD, Verhaegen M (2006) Development of advanced driver assistance systems with vehicle hardware-in-the-loop simulations. Veh Syst Dyn 44(7):569–590

Mihály A, Baranyi M, Németh B, Gáspár P (2017) Tuning of look-ahead cruise control in hil vehicle simulator. Period Polytech Transp Eng 45:157–161

Rajamani R (2005) Vehicle dynamics and control. Springer

Vandi G, Cavina N, Corti E, Mancini G, Moro D, Ponti F, Ravaglioli V (2014) Development of a software in the loop environment for automotive powertrain systems. Energy Procedia. In: ATI 2013-68th conference of the Italian thermal machines engineering association 45:789–798

Appendix A
Brief Summary of the Model-Based Robust LPV Control Design

A.1 Control-Oriented Modeling of LPV Systems

In the model-based analysis and synthesis, an LPV (Linear Parameter Varying) model structure is often applied. In the state space representation of LPV model, the system matrices $A(\rho)$, $B_1(\rho)$ and $B_2(\rho)$ contain the scheduling variable ρ:

$$\dot{x} = A(\rho)x + B_1(\rho)d + B_2(\rho)u, \tag{A.1}$$

where x is the vector of the state variables, d is the disturbance, and u is the control input. Note that if ρ is constant, the model structure leads to an LTI (Linear Time Invariant):

$$\dot{x} = Ax + B_1d + B_2u \tag{A.2}$$

where A, B_1, and B_2 are constant matrices. Moreover, if $\rho = \rho(x)$ depends on the state variables, the model structure leads to qLPV (quasi-Linear Parameter Varying) model structure. It is important to emphasize that the LPV and qLPV models are able to handle both the uncertainties and the nonlinearities.

The LPV modeling is based on the possibility of rewriting the plant in a form in which nonlinear terms can be hidden with suitably defined scheduling variables, see Shamma and Athans (1991), Rough and Shamma (2000). The advantage of LPV models is that the entire operational intervals of nonlinear systems can be defined and a well-developed linear system theory to analyze and design nonlinear systems can be used. The scheduling variable is measured directly, or can be calculated by using the measured (or estimated) signals. The LPV model representations are not unique and determining a suitable representation depending on the intended scheduling variables is not a trivial task. Note that the classical linearization procedure leads to a linear model, which is valid only in the equilibrium points.

P. Gáspár and B. Németh, *Predictive Cruise Control for Road Vehicles Using Road and Traffic Information*, Advances in Industrial Control,
https://doi.org/10.1007/978-3-030-04116-8

In the control design, the state-space equation is augmented by the performance equations z and the equations of the measured signals y.

$$z = C_1(\rho)x + D_{11}(\rho)d + D_{12}(\rho)u, \tag{A.3}$$
$$y = C_2(\rho)x + D_{21}(\rho)d + D_{22}(\rho)u, \tag{A.4}$$

where C_1, C_2, D_{11}, D_{12}, D_{21}, D_{22} are parameter-dependent matrices.

In the construction of the closed-loop interconnection structure, weighting functions are designed in order to create a trade-off between performance specifications. They can be considered as penalty functions, i.e., weights should be large in a frequency range where small signals are desired and small where large performance outputs can be tolerated. Moreover, weighting functions may be used to reflect some restrictions on the actuator and on the input/output signals.

The deviations between the plant and its model are described by uncertainty structure denoted by Δ. This leads to the $P - \Delta$ structure, in which P is the model, and weighting functions W_p are applied to the performances. An unstructured uncertainty Δ_m is modeled by connecting an unknown but bounded perturbation to the plant. This type of uncertainty structure requires little information, i.e., only the bound and the type of connection. Structured uncertainties arise when the plant is subject to multiple perturbations. Multiple perturbations occur when the plant contains a number of uncertain parameters or when the plant contains multiple unstructured uncertainties. In practice, many systems involve parametric uncertainties Δ_r, which are defined by their operation interval.

The detailed description of the $P - \Delta$ structure is shown in Fig. A.1a, while the compact form is shown in Fig. A.1b. The figures also contain the external disturbances and the noises with weighting functions W_w and W_n.

The robust control design is based on the so-called $P - K - \Delta$ structure, in which K is the control. This structure is shown in Fig. A.2a, while the compact form is shown in Fig. A.2b. The purpose of the design is to construct the control K, which generates the control input by using the measured signals and keeps the operator norm of the impacts of the disturbances on the performances within a predefined value γ. In the calculation, the $\mathscr{P}$ operation range of the scheduling variables ρ is taken into consideration.

$$\inf_K \sup_{\rho \in \mathscr{P}} \sup_{\|d\|_2 \neq 0, d \in \mathscr{L}_2} \frac{\|z\|_2}{\|d\|_2} < \gamma \tag{A.5}$$

A.2 Gain-Scheduled $\mathscr{H}_\infty$ Controllers

The theoretical and practical approaches of gain-scheduled $\mathscr{H}_\infty$ controllers usually lead to an LMI-based $\mathscr{H}_\infty$ synthesis, see Apkarian and Gahinet (1995), Gahinet and Apkarian (1994), Gahinet and Apkarian (1996). An LPV system P, which maps an exogenous input d and a control input u to a controlled output z and a measured

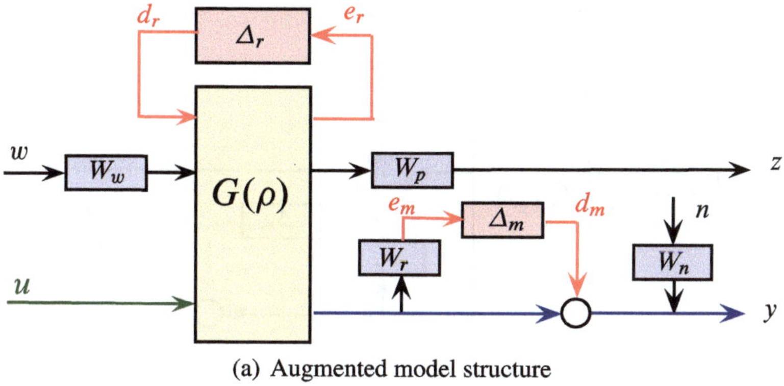

Fig. A.1 $P - \Delta$ structure with the weighting functions

output y, in the following state-space representation way:

$$\dot{x} = Ax + B_1 d + B_2 u, \tag{A.6}$$

$$z = C_1 x + D_{12} u, \tag{A.7}$$

$$y = C_2 x + D_{21} d. \tag{A.8}$$

The purpose of the control design problem is to find an LPV control structure K such that the closed-loop system is internally stable for all parameter trajectories with $\gamma^2 \Theta^T \Theta \leq 1$ and the induced $\mathscr{L}_2$-norm of the operator $T(P, K, \Theta)$ satisfies

$$\max_{\|\Theta\|_\infty \leq 1/\gamma} \|T(P, K, \Theta)\|_\infty < \gamma, \tag{A.9}$$

where γ is a prescribed performance level.

The LPV problem can be interpreted as a robust performance problem for the nominal linear time invariant (LTI) plant P_a in the face of the norm-bounded uncertainty $\Delta \oplus \Delta$. The structure of the uncertainty is the following: $\Delta = diag(\theta_1 I_{r1}, ..., \theta_k I_{rk})$, where $r_i > 1$ whenever the parameter θ_i is repeated. Consider the set of positive-definite similarity scalings associated with Δ:

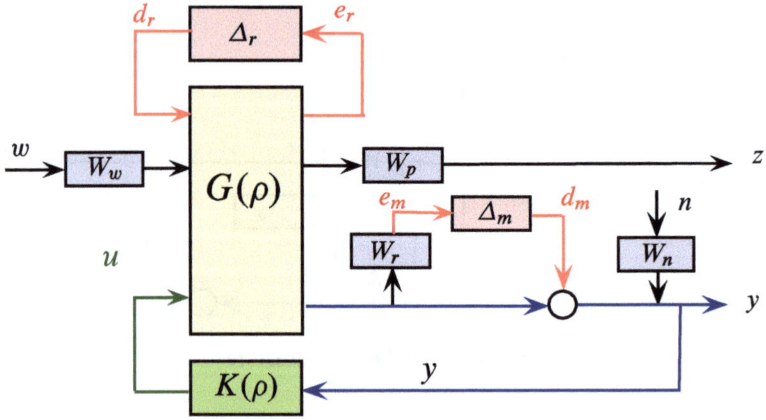

(a) Augmented model structure with control

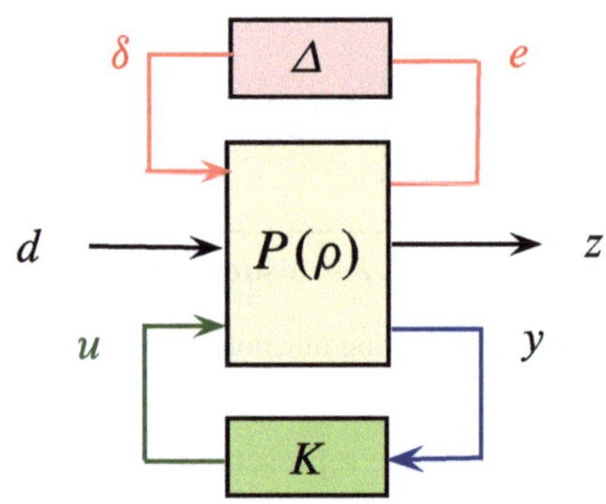

(b) $P-K-\Delta$ struktúra

Fig. A.2 $P - K - \Delta$ structure with the weighting functions and the control K

$$L_\Delta = \{L > 0 : L\Theta = \Theta L, \forall \Theta \in \Delta\} \tag{A.10}$$

Given L_Δ, the set of scalings commuting with the repeated structure $\Delta \oplus \Delta$ is readily deduced as

$$L_{\Delta\oplus\Delta} = \left\{ \begin{bmatrix} L_1 & L_2 \\ L_2^T & L_3 \end{bmatrix} > 0 : L_1, L_3 \in L_\Delta, L_2\Theta = \Theta L_2, \forall \Theta \in \Delta \right\} \tag{A.11}$$

From small gain theory, a sufficient condition for robust performance in the face of the uncertainty $\Delta \oplus \Delta$, or equivalently for the existence of gain-scheduled controllers is as follows:

Theorem A.1 (γ-suboptimal gain-scheduled $\mathscr{H}_\infty$ controller) *Consider an uncertainty structure Δ and the associated set of similarity scalings $L_{\Delta\oplus\Delta}$ defined in (A.11). If there exist a scaling matrix $L \in L_{\Delta\oplus\Delta}$ and an LTI control structure K such* *that the nominal closed-loop system $F_\ell(P_a, K)$ is internally stable and satisfies the following inequality*

$$\left\| \begin{bmatrix} L^{1/2} & 0 \\ 0 & I \end{bmatrix} F_\ell(P_a, K) \begin{bmatrix} L^{-1/2} & 0 \\ 0 & I \end{bmatrix} \right\|_\infty < \gamma, \tag{A.12}$$

then $F_\ell(K, \Theta)$ is a γ-suboptimal gain-scheduled $\mathscr{H}_\infty$ controller. See Apkarian and Gahinet (1995).

The LMI approach applied for gain-scheduled $\mathscr{H}_\infty$ controller is based on the Scaled Bounded Real Lemma, which is formalized in the following.

Theorem A.2 (Scaled Bounded Real Lemma) *Consider a parameter structure Δ, the associated scaling set L_Δ defined in (A.10), and closed-loop system. The following statements are equivalent.*

- *A is stable and there exists $L \in L_\Delta$ such that*

$$\left\| L^{1/2}(D + C(sI - A)^{-1}B)L^{-1/2} \right\|_\infty < \gamma \tag{A.13}$$

- *There exist positive-definite solutions P and $L \in L_\Delta$ to the matrix inequality*

$$\begin{bmatrix} A^T P + PA & PB & C^T \\ B^T P & -\gamma L & D^T \\ C & D & -\gamma L^{-1} \end{bmatrix} < 0. \tag{A.14}$$

See Apkarian and Gahinet (1995).

Using Theorem A.1, the solution of the scaled $\mathscr{H}_\infty$ controller problem is presented. In the next theorem, the scaled $\mathscr{H}_\infty$ problem is considered as LTI plants with arbitrary uncertainty structures Δ. The general statement of scaled $\mathscr{H}_\infty$ problems is as follows. Given $\gamma > 0$, an uncertainty structure Δ, and the associated scaling set L_Δ. Find $L \in L_\Delta$ and an LTI controller K such that the closed-loop system is internally stable and

$$\left\| L^{1/2} F_\ell(G, K) L^{-1/2} \right\|_\infty < \gamma \tag{A.15}$$

Theorem A.3 (Solution of the scaled $\mathscr{H}_\infty$ controller problem) *Consider a plant G with uncertainty structures Δ and an associated scaling set L_Δ. Let N_X and N_Y denote orthonormal bases of the null spaces $(B_2^T, D_{12}^T, 0)$ and $(C_2, D_{21}, 0)$, respectively. The suboptimal scaled $\mathscr{H}_\infty$ problem is solvable if and only if there exist pairs of symmetric matrices $X, Y \in \mathscr{R}^{n\times n}$ and $L, J \in \mathscr{R}^{r\times r}$ with $L \in L_\Delta$, $J \in J_\Delta$ satisfying the following system of LMIs:*

$$N_X^T \begin{bmatrix} AX + XA^T & XC_1^T & B_1 \\ C_1 X & -\gamma J & 0 \\ B_1^T & 0 & -\gamma L \end{bmatrix} N_X < 0, \tag{A.16}$$

$$N_Y^T \begin{bmatrix} A^T Y + YA & YB_1 & C_1^T \\ B_1^T Y & -\gamma L & 0 \\ C_1 & 0 & -\gamma J \end{bmatrix} N_Y < 0, \tag{A.17}$$

$$\begin{bmatrix} X & I \\ I & Y \end{bmatrix} \geq 0, \tag{A.18}$$

$$LJ = I. \tag{A.19}$$

See Apkarian and Gahinet (1995).

Inequalities (A.16)–(A.18) are LMIs in R, S, L, J and the structure constraints $L \in L_\Delta$ and $J \in J_\Delta$ are convex constraints. However, constraint (A.19) is strongly non-convex and thus its numerical tractability is unclear. Fortunately, in the gain-scheduled $\mathscr{H}_\infty$ control problem, the non-convex constraint $LJ = I$ disappears. As a result, a sufficient condition can be obtained that is pure LMI and is therefore tractable. The theorem, which is used in the gain-scheduled control design, is the following.

Theorem A.4 (Existence of the gain-scheduled $\mathscr{H}_\infty$ controller problem) *Consider an LPV plant given by the LFT interconnection structure in which P is a proper LTI plant and Θ is the parameter operator. Let Δ denote the structure set associated with Θ and L_Δ denote the corresponding set of scaling matrices. Let N_X and N_Y denote orthonormal bases of the null spaces $(B_2^T, D_{\theta 2}^T, D_{12}^T, 0)$ and $(C_2, D_{2\theta}, D_{21}, 0)$, respectively. The gain-scheduled $\mathscr{H}_\infty$ control problem is solvable if and only if there exist pairs of symmetric matrices $X, Y \in \mathscr{R}^{n \times n}$ and $L, J \in \mathscr{R}^{r \times r}$ with $L \in L_\Delta$, $J \in J_\Delta$ satisfying the following system of $LMIs$:*

$$N_X^T \begin{bmatrix} AX + XA^T & \begin{bmatrix} XC_\theta & XC_1 \end{bmatrix} & \begin{bmatrix} B_\theta & B_1 \end{bmatrix} \\ \begin{bmatrix} C_\theta X \\ C_1 X \end{bmatrix} & -\gamma \begin{bmatrix} J & 0 \\ 0 & I \end{bmatrix} & \begin{bmatrix} 0 & 0 \\ 0 & 0 \end{bmatrix} \\ \begin{bmatrix} B_\theta \\ B_1 \end{bmatrix} & \begin{bmatrix} 0 & 0 \\ 0 & 0 \end{bmatrix} & -\gamma \begin{bmatrix} L & 0 \\ 0 & I \end{bmatrix} \end{bmatrix} N_X < 0, \tag{A.20}$$

$$N_Y^T \begin{bmatrix} A^T Y + YA & \begin{bmatrix} YB_\theta & YB_1 \end{bmatrix} & \begin{bmatrix} C_\theta & C_1 \end{bmatrix} \\ \begin{bmatrix} B_\theta Y \\ B_1 Y \end{bmatrix} & -\gamma \begin{bmatrix} L & 0 \\ 0 & I \end{bmatrix} & \begin{bmatrix} 0 & 0 \\ 0 & 0 \end{bmatrix} \\ \begin{bmatrix} C_\theta \\ C_1 \end{bmatrix} & \begin{bmatrix} 0 & 0 \\ 0 & 0 \end{bmatrix} & -\gamma \begin{bmatrix} J & 0 \\ 0 & I \end{bmatrix} \end{bmatrix} N_Y < 0, \tag{A.21}$$

$$\begin{bmatrix} X & I \\ I & Y \end{bmatrix} \geq 0, \tag{A.22}$$

$$\begin{bmatrix} L & I \\ I & J \end{bmatrix} \geq I. \tag{A.23}$$

See Apkarian and Gahinet (1995).

Theorem A.4 provides sufficient conditions for the existence of gain-scheduled $\mathcal{H}_\infty$ controllers. An algorithm for the computation of the components of the controller is proposed by Apkarian and Gahinet (1995). Pairs (X, Y) are called feasible for the LMI system and computing feasible pairs is a convex optimization problem. Efficient polynomial time algorithms are available to solve this LMI feasibility problem, see Nesterov and Nemirovsky (1993), Nemirovsky and Gahinet (1994).

A.3 Analysis of LPV Systems

Guaranteeing stability and performance requirements is a central question in the control design, Bokor and Balas (2005). Let us see the following LPV systems:

$$\begin{bmatrix} \dot{x} \\ z \end{bmatrix} = \begin{bmatrix} A(\rho) & B(\rho) \\ C(\rho) & D(\rho) \end{bmatrix} \begin{bmatrix} x \\ d \end{bmatrix} \tag{A.24}$$

In practice, the analysis and synthesis based on a single Lyapunov function (SLF) approach are often applied. In this case, it is assumed that the variation of the scheduling variables can be arbitrarily fast. The system meets the quadratic stability if there exists a quadratic Lyapunov function $V(x(t)) = x^T(t)Px(t) > 0$ such that $P = P^T > 0$ is constant and the following linear matrix inequality (LMI) fulfills, see Becker et al. (1993), Scherer et al. (1997):

$$A^T(\rho)P + PA(\rho) < 0, \quad \forall \rho \in \mathscr{P}, \tag{A.25}$$

The performances of the system are the following. If there exists constant $P = P^T > 0$ with the following LMI:

$$\begin{bmatrix} A^T(\rho)P + PA(\rho) & PB(\rho) & C^T(\rho) \\ B^T(\rho)P & -\gamma I & D^T(\rho) \\ C^T(\rho) & D(\rho) & -\gamma I \end{bmatrix} < 0, \quad \forall \rho \in \mathscr{P}, \tag{A.26}$$

then the $\mathscr{L}_2$ gain of LPV system is less than the predefined γ with zero initial conditions $x(0) = 0$.

The above stability and performance conditions often lead to conservative solutions. Moreover, they are not met numerically even for stable systems. To reduce the conservatism, the P matrix is selected parameter-dependent matrix instead of constant one. Thus, a parameter-dependent Lyapunov function (PDLF) is applied. The motivation reason for that is that when the bandwidth of the actuators or the signals is disregarded, it leads to an infinite rate bound on the scheduling variable in an impractical way. If the rate on ρ is assumed to be bounded, a less conservative result for the class of systems is yielded. This is $|\dot{\rho}_i| \le \beta_i, i \in \{1, s\}$.

The LPV system with parameter-dependent scheduling variables is as follows:

$$\begin{bmatrix} \dot{x} \\ z \end{bmatrix} = \begin{bmatrix} A(\rho, \dot{\rho}) & B(\rho, \dot{\rho}) \\ C(\rho, \dot{\rho}) & D(\rho, \dot{\rho}) \end{bmatrix} \begin{bmatrix} x \\ d \end{bmatrix}, \tag{A.27}$$

for all ρ and $\beta_i \leq \nu(i$

The stability theorem is the following. If a function A is parametrically dependent stable over the compact set $\mathscr{P}$, then there exists some $\delta > 0$ and a parameter-dependent matrix $P(\rho) = P^T(\rho) > 0$ such that

$$A^T(\rho, \dot{\rho})P(\rho) + P(\rho)A(\rho, \dot{\rho}) + \frac{dP}{dt} < -\delta I_n, \quad \forall \rho \in \mathscr{P} \tag{A.28}$$

For an LPV system if function A is parametrically dependent stable, then the system is a parametrically dependent stable LPV system.

The following theorem formulates a sufficient condition to test whether the induced $\mathscr{L}_2$-norm of an LPV system is less than a prescribed performance level $\gamma > 0$, see Packard and Becker (1992), Wu et al. (1996), Wu (2001). Given a compact $\mathscr{P}$, finite nonnegative numbers $\{\nu_i\}_{i=1}^{s}$. If there exists a function $\mathscr{P}$ such that $P(\rho) > 0$ and

$$\begin{bmatrix} A^T(\rho, \beta)P(\rho) + P(\rho)A(\rho, \beta) + \sum_{i=1}^{s}\left(\beta_i \frac{\partial P}{\partial \rho_i}\right) & P(\rho)B(\rho, \beta) & C^T(\rho, \beta) \\ B^T(\rho, \beta)P(\rho) & -\gamma I_{n_d} & D^T(\rho, \beta) \\ C(\rho, \beta) & D(\rho, \beta) & -\gamma I_{n_c} \end{bmatrix} < 0, \tag{A.29}$$

for all $\rho \in \mathscr{P}$ and $|\beta_i| \leq \nu_i$, $i \in \{1, s\}$, then the function A is parametrically dependent stable over $\mathscr{P}$ and there exists a scalar δ with $0 \leq \delta < \gamma$ such that $\|G_{\mathscr{F}^{\nu}_{\mathscr{P}}}\|_{i,2} \leq \delta$.

The condition of the theorem is an infinite-dimensional convex problem. By approximating the function space with finite basis functions, we can simplify the condition to a finite-dimensional convex problem and solve the analysis problem using efficient convex optimization techniques. In practice, the gridding of parameter space can lead to computationally less expensive task, see Wu et al. (1996).

In the parameter-dependent methods, the theorem related to stability and performances uses more complex LMIs, since besides P also its time derivative dP/dt also occurs. However, the parameter-dependent representation has advantage in both the design of the control and its implementation, since the conservatism of the controller is significantly reduced.

A.4 Synthesis of LPV Systems

In the following, the parameter-dependent output-feedback control problem for LPV systems with bounded parameter variation rates is studied. This problem determines the existence of a parameter-dependent controller which parametrically dependent stabilizes the closed-loop LPV system and guarantees that the induced $\mathscr{L}_2$-norm of the closed-loop system less than γ. The derivative of parameter is assumed to be achievable (or measurable) in real time to construct such a controller.

A compact set $\mathscr{P} \subset \mathscr{R}^S$ is given. Consider the open-loop LPV system in the following way:

$$\dot{x} = A(\rho)x + B_1(\rho)d + B_2(\rho)u, \tag{A.30}$$

$$z = C_1(\rho)x + D_{11}(\rho)d + D_{12}(\rho)u, \tag{A.31}$$

$$y = C_2(\rho)x + D_{21}(\rho)d + D_{22}(\rho)u, \tag{A.32}$$

where $\rho \in \mathscr{F}^{\nu}_{\mathscr{P}}$, $x \in \mathscr{R}^n$, $d \in \mathscr{R}^{n_d}$, $z \in \mathscr{R}^{n_z}$, $u \in \mathscr{R}^{n_u}$, $y \in \mathscr{R}^{n_y}$. The following assumptions for the generalized plant are made. The assumption $D_{22}(\rho) = 0$ can be relaxed by including a feed through term to the controller for the modified plant. $D_{12}(\rho)$ is full column rank for all $\rho \in \mathscr{P}$. $D_{21}(\rho)$ is full row rank for all $\rho \in \mathscr{P}$. The relaxation of these assumptions leads to singular $\mathscr{H}_\infty$ problem.

A compact set $\mathscr{P} \subset \mathscr{R}^S$, an integer $m \geq 0$, and the parametrically dependent m-dimensional linear feedback controller with the continuous functions $A_k : \mathscr{R}^S \to \mathscr{R}^{m \times m}$, $B_k : \mathscr{R}^S \to \mathscr{R}^{m \times n_y}$, $C : \mathscr{R}^S \to \mathscr{R}^{n_u \times m}$, $D : \mathscr{R}^S \to \mathscr{R}^{n_u \times n_y}$ are given. The controller depends on the parameter and its derivative according to the following equation:

$$\dot{x}_k = A_k(\rho, \dot{\rho})x_k + B_k(\rho, \dot{\rho})y, \tag{A.33}$$

$$u_k = C_k(\rho, \dot{\rho})x_k + D_k(\rho, \dot{\rho})y, \tag{A.34}$$

where $\rho \in \mathscr{F}^{\nu}_{\mathscr{P}}$, x_k is the m-dimensional controller states.

In the following, the parameter-dependent output-feedback controller synthesis is presented. In order to achieve simple formulas, $D_{11} = 0$ is assumed. However, the results can be generalized to $D_{11} \neq 0$ case. The following simplifications are also made: $D_{12} = \begin{bmatrix} 0 & I_{nz_2} \end{bmatrix}$ and $D_{21} = \begin{bmatrix} I_{nd_2} & 0 \end{bmatrix}$

The open-loop system is modified in the following form:

$$\begin{bmatrix} \dot{x}(t) \\ \hline z_1(t) \\ z_2(t) \\ y(t) \end{bmatrix} = \left[\begin{array}{c|ccc} A(\rho(t)) & B_{11} & B_{12} & B_2 \\ \hline C_{11} & 0 & 0 & 0 \\ C_{12} & 0 & 0 & I_{nz_2} \\ C_2 & 0 & I_{nd_2} & 0 \end{array}\right] \begin{bmatrix} x(t) \\ \hline d_1(t) \\ d_2(t) \\ u_c(t) \end{bmatrix}, \tag{A.35}$$

where $\rho \in \mathscr{F}^{\nu}_{\mathscr{P}}$, $d_1 \in \mathscr{R}^{n_{d1}}$, $d_2 \in \mathscr{R}^{n_{d2}}$, $z_1 \in \mathscr{R}^{n_{e1}}$, and $z_2 \in \mathscr{R}^{n_{e2}}$.

The closed-loop system with $\dot{x}_c^T(t) = \begin{bmatrix} x^T(t) & x_k^T(t) \end{bmatrix}$, $z^T(t) = \begin{bmatrix} z_1^T(t) & z_2^T(t) \end{bmatrix}$, $d^T(t) = \begin{bmatrix} d_1^T(t) & d_2^T(t) \end{bmatrix}$ is given by

$$\dot{x}_c = A_c(\rho, \dot{\rho})x_c + B_c(\rho, \dot{\rho})d, \quad (A.36)$$

$$z = C_c(\rho, \dot{\rho})x_c + D_c(\rho, \dot{\rho})d, \quad (A.37)$$

where

$$A_c(\rho, \dot{\rho}) = \begin{bmatrix} A(\rho) + B_2(\rho)D_K(\rho, \dot{\rho})C_2(\rho) & B_2(\rho)C_K(\rho, \dot{\rho}) \\ B_K(\rho, \dot{\rho})C_2(\rho) & A_K(\rho, \dot{\rho}) \end{bmatrix}, \quad (A.38)$$

$$B_c(\rho, \dot{\rho}) = \begin{bmatrix} B_{11}(\rho) & B_{12}(\rho) + B_2(\rho)D_K(\rho, \dot{\rho}) \\ 0 & B_K(\rho, \dot{\rho}) \end{bmatrix}, \quad (A.39)$$

$$C_c(\rho, \dot{\rho}) = \begin{bmatrix} C_{11}(\rho) & 0 \\ C_{12}(\rho) + D_K(\rho, \dot{\rho})C_2(\rho) & C_K(\rho, \dot{\rho}) \end{bmatrix}, \quad (A.40)$$

$$D_c(\rho, \dot{\rho}) = \begin{bmatrix} 0 & 0 \\ 0 & D_K(\rho, \dot{\rho}) \end{bmatrix}. \quad (A.41)$$

Definition A.1 (*The parameter-dependent γ performance problem*) An open-loop LPV system and the performance level $\gamma > 0$ are given. The parameter-dependent γ performance problem is solvable if there exist an integer $m \geq 0$, a function $P : \mathscr{R}^S \to S^{(n+m)\times(m+n)}$, and continuous matrix functions $A_k : \mathscr{R}^S \to \mathscr{R}^{m\times m}$, $B_k : \mathscr{R}^S \to \mathscr{R}^{m\times n_y}$, $C_k : \mathscr{R}^S \to \mathscr{R}^{n_u\times m}$, $D_k : \mathscr{R}^S \to \mathscr{R}^{n_u\times n_y}$ such that $P(\rho) > 0$ and

$$\begin{bmatrix} A_c^T(\rho, \beta)P(\rho) + P(\rho)A_c(\rho, \beta) + \sum_{i=1}^{s} \beta_i \frac{\partial P}{\partial \rho_i} & P(\rho)B_c(\rho, \beta) & \gamma^{-1}C_c^T(\rho, \beta) \\ B_c^T(\rho, \beta)P(\rho) & -I_{n_d} & \gamma^{-1}D_c^T(\rho, \beta) \\ \gamma^{-1}C_c^T(\rho, \beta) & \gamma^{-1}D_c^T(\rho, \beta) & -I_{n_z} \end{bmatrix} < 0, \quad (A.42)$$

for all $\rho \in \mathscr{P}$ and $|\beta_i| \leq \nu_i$, $i = 1, 2 \ldots, s$. The closed-loop matrices are defined in (A.38).

This problem is a generalization of the standard suboptimal $\mathscr{H}_\infty$ control problem. The output-feedback controller synthesis is based on the following theorem.

Theorem A.5 *A compact set $\mathscr{P}$, the performance level $\gamma > 0$, and the LPV system are given. The parameter-dependent γ-performance problem is solvable if and only if there exist matrix functions $X : \mathscr{R}^S \to \mathscr{R}^{n\times n}$, and $Y : \mathscr{R}^S \to \mathscr{R}^{n\times n}$, such that for all $\rho \in \mathscr{P}$, $X(\rho) > 0$, $Y(\rho) > 0$, and*

$$\begin{bmatrix} X(\rho)\hat{A}^T(\rho) + \hat{A}(\rho)X(\rho) - \sum_{i=1}^{s} \pm\left(\nu_i \frac{\partial X}{\partial \rho_i}\right) - B_2(\rho)B_2^T(\rho) & X(\rho)C_{11}^T(\rho) & \gamma^{-1}B_1(\rho) \\ C_{11}(\rho)X(\rho) & -I_{n_{z1}} & 0 \\ \gamma^{-1}B_1^T(\rho) & 0 & -I_{n_d} \end{bmatrix} < 0, \tag{A.43}$$

$$\begin{bmatrix} \tilde{A}^T(\rho)Y(\rho) + Y(\rho)\tilde{A}(\rho) + \sum_{i=1}^{s} \pm\left(\nu_i \frac{\partial Y}{\partial \rho_i}\right) - C_2^T(\rho)C_2(\rho) & Y(\rho)B_{11}(\rho) & \gamma^{-1}C_1^T(\rho) \\ B_{11}^T(\rho)Y(\rho) & -I_{n_{d1}} & 0 \\ \gamma^{-1}C_1(\rho) & 0 & -I_{n_z} \end{bmatrix} < 0, \tag{A.44}$$

$$\begin{bmatrix} X(\rho) & \gamma^{-1}I_n \\ \gamma^{-1}I_n & Y(\rho) \end{bmatrix} \geq 0. \tag{A.45}$$

where

$$\hat{A}(\rho) = A(\rho) - B_2(\rho)C_{12}(\rho), \tag{A.46}$$

$$\tilde{A}(\rho) = A(\rho) - B_{12}(\rho)C_2(\rho). \tag{A.47}$$

If the conditions are satisfied, it is possible—by continuity and compactness—to perturb $X(\rho)$ such that the two LMIs (A.43) and (A.44) still hold and $Q(\rho) = Y(\rho) - \gamma^{-2}X^{-1}(\rho) > 0$ uniformly on $\mathscr{P}$. Define now the following variables:

$$F(\rho) = -[B_2(\rho)X^{-1}(\rho) + D_{12}^T C_1(\rho)], \tag{A.48}$$

$$L(\rho) = -[Y^{-1}(\rho)C_2^T(\rho) + B_1(\rho)D_{21}^T], \tag{A.49}$$

$$H(\rho,\dot{\rho}) = -[X^{-1}(\rho)A_F(\rho) + A_F^T(\rho)X^{-1}(\rho) + \sum_{i=1}^{s}\left(\dot{\rho}\frac{\partial X^{-1}}{\partial \rho_i}\right) + C_F^T(\rho)C_F(\rho) + \gamma^{-2}X^{-1}(\rho)B_1(\rho)B_1^T(\rho)X^{-1}(\rho)], \tag{A.50}$$

with

$$A_F(\rho) = A(\rho) + B_2(\rho)F(\rho), \tag{A.51}$$

$$C_F(\rho) = C_1(\rho) + D_{12}(\rho)F(\rho). \tag{A.52}$$

Furthermore,

$$M(\rho,\dot{\rho}) = H(\rho,\dot{\rho}) + \gamma^2 Q(\rho)[-Q^{-1}(\rho)Y(\rho)L(\rho)D_{21} - B_1(\rho)]B_1^T(\rho)X^{-1}(\rho). \tag{A.53}$$

The strictly proper controller that solves the feedback problem is given by

$$A_k(\rho, \dot{\rho}) = A(\rho) + B_2(\rho)F(\rho) + Q^{-1}(\rho)Y(\rho)L(\rho)C_2(\rho) - \gamma^{-2}Q^{-1}(\rho)M(\rho, \dot{\rho}), \tag{A.54}$$

$$B_k(\rho) = -Q^{-1}Y(\rho)L(\rho), \tag{A.55}$$

$$C_k(\rho) = F(\rho), \tag{A.56}$$

$$D_k(\rho) = 0. \tag{A.57}$$

See Wu (1995).

The proof of this theorem and the discussion of the general case $D_{11}(\rho) \neq 0$ are also in Wu (1995). Equation (A.43) is for the state feedback, (A.44) is for the output estimation and (A.45) is the coupling condition. The LMIs in Theorem A.5 lead to an infinite-dimensional convex feasibility problem.

Note that the controller dynamics matrix $A_k(\rho, \dot{\rho})$ depends explicitly on $\dot{\rho}$. In order to construct a parameter-dependent controller, both ρ and $\dot{\rho}$ must be measured or available. In case of a single scheduling variable, i.e., when $\dot{\rho}$ is not measured, it is possible to perform a ρ-dependent change of variables to remove $\dot{\rho}$ dependence as follows.

In the scalar parameter case, the controller dynamics from Theorem A.5 take the form

$$\dot{x}_k = A_1(\rho)x_k + \dot{\rho}A_2(\rho)x_k + B_k(\rho). \tag{A.58}$$

Now consider the change of variables

$$x_{k,new} = T(\rho)x_k + \dot{\rho}A_2(\rho)x_k + B_k(\rho), \tag{A.59}$$

where $T(\rho)$ is invertible. It is easy to show that if $T(\rho)$ satisfies the linear matrix differential equation

$$\frac{dT(\rho)}{d\rho} = -T(\rho)A_2(\rho), \tag{A.60}$$

then the resulting controller dynamics in the new coordinates $x_{k,new}$ will be independent of $\dot{\rho}$. In case of multiple scheduling variables, a similar method can be applied. However, the analytical solvability is more difficult. More details, see Packard and Balas (1997).

An alternative solution is to simply restrict the choice X to be independent of $\dot{\rho}$ in Eq. (A.50). However, this approach may be conservative, since the rate of change of the parameters is not bounded. Another possible solution is to apply a suitable extrapolation algorithm in order to achieve an estimation of the parameter $\dot{\rho}$. The disadvantage of this approach is that the sources of the scheduling variables are not independent.

In the following, the infinite convex feasibility conditions in Theorem A.5 will be converted to finite-dimensional LMIs. In the solution, a finite number of basis functions are selected to parameterize infinite-dimensional function space.

Theorem A.6 *Given a finite number of scalar, continuously differentiable functions* $\{f_i\}_{i=1}^N$ *and* $\{g_i\}_{i=1}^N$ *with the parametrization:*

$$X(\rho) = \sum_{i=1}^{N} f_i(\rho) X_i, \tag{A.61}$$

$$Y(\rho) = \sum_{i=1}^{N} g_i(\rho) Y_i. \tag{A.62}$$

The parameter-dependent γ*-performance problem is solvable if there exist matrices* $\{X_i\}_{i=1}^N$, $X_i \in \mathscr{S}^{n\times n}$, *and* $\{Y_i\}_{i=1}^N$, $Y_i \in \mathscr{S}^{n\times n}$ *such that for all* $\rho \in \mathscr{P}$, $X(\rho) > 0$, $Y(\rho) > 0$, *and*

$$\begin{bmatrix} \sum_{i=1}^{N} f_i(\rho)\left(X_i \hat{A}^T(\rho) + \hat{A}(\rho) X_i\right) - \sum_{j=1}^{s} \pm\left(\nu_j \sum_{i=1}^{N} \frac{\partial f_i}{\partial \rho_i} X_i\right) - B_2(\rho) B_2^T(\rho) & \sum_{i=1}^{N} f_i(\rho) X_i C_{11}^T(\rho) & \gamma^{-1} B_1(\rho) \\ C_{11}(\rho) \sum_{i=1}^{N} f_i(\rho) X_i & -I_{n_{z1}} & 0 \\ \gamma^{-1} B_1^T(\rho) & 0 & -I_{n_d} \end{bmatrix} < 0, \tag{A.63}$$

$$\begin{bmatrix} \sum_{i=1}^{N} g_i(\rho)\left(\tilde{A}^T(\rho) Y_i + Y_i \tilde{A}(\rho)\right) + \sum_{j=1}^{s} \pm\left(\nu_j \sum_{i=1}^{N} \frac{\partial g_i}{\partial \rho_i} Y_i\right) - C_2^T(\rho) C_2(\rho) & \sum_{i=1}^{N} g_i(\rho) Y_i B_{11}(\rho) & \gamma^{-1} C_1^T(\rho) \\ B_{11}^T(\rho) \sum_{i=1}^{N} g_i(\rho) Y_i & -I_{n_{d1}} & 0 \\ \gamma^{-1} C_1(\rho) & 0 & -I_{n_z} \end{bmatrix} < 0, \tag{A.64}$$

$$\begin{bmatrix} \sum_{i=1}^{N} f_i(\rho) X_i & \gamma^{-1} I_n \\ \gamma^{-1} I_n & \sum_{i=1}^{N} g_i(\rho) Y_i \end{bmatrix} \geq 0. \tag{A.65}$$

where the matrices $\hat{A}(\rho)$ *and* $\tilde{A}(\rho)$ *are defined in Eq. (A.46). See Wu (1995).*

The other problem is that the inequalities must hold for all $\rho \in \mathscr{P}$, which entails that an infinite number of constraints must be checked. In order to solve this infinite constraints convex problem, the compact set $\mathscr{P}$ must be gridded. For example, if a hyper rectangle $\mathscr{P} \subset \mathscr{R}^S$ is gridded with L points in each dimension, then the convex problem to determine appropriate $\{X_i\}_{i=1}^N$ and $\{Y_i\}_{i=1}^N$ includes approximately $L^s(2^{s+1} + 1)$ LMIs. Another problem is the complete lack of guidance provided by the theory to pick the basis functions, namely f_i and g_i. Thus, usually intuitive solutions for basis functions are applied.

A.5 Formulation of a Nonlinear Controller

When the LPV controller has been synthesized, the relation between the state, or output, and the parameter $\rho = \sigma(x)$ is used in the LPV controller, such that a nonlinear controller is obtained:

$$\dot{x}_c = A_c(\sigma(x))x_c + B_c(\sigma(x))y, \tag{A.66}$$

$$u = C_c(\sigma(x))x_c + D_c(\sigma(x))y. \tag{A.67}$$

Note that it is assumed that $\sigma(x)$ is measured or depends only on measured signals. According to the properties of the LPV description, the LPV system with $\rho = \sigma(x)$ is equal to the nonlinear model.

The LPV design is performed in such a way that the closed-loop system specifications hold for all parameter values and parameter derivative values in the bounding boxes, i.e., $\rho \in \Omega$ and $\dot{\rho} \in \tilde{\Omega}$. One trajectory of the parameter corresponds to the trajectory given by the parameter to state relation of the LPV description. Hence, the LPV controller meets the closed-loop specifications for the nonlinear system, as long as the parameter trajectories remain in the bounded boxes. The LPV control guarantees to meet closed-loop system specifications in the validity domain even under transients.

Appendix B
Brief Summary of the Maximum Controlled Invariant Sets

B.1 Fundamentals of SOS Programming Technique

n the following, the fundamental concepts concerning the SOS programming method are summarized. The method is suitable for analysis and control nonlinear polynomial systems.

Several papers deal with SOS programming, which has been elaborated in the recent decade for control purposes. It is an efficient tool for finding feasible solutions to polynomial inequalities. In SOS programming, this problem is transformed into a semi-definite optimization task. Important theorems in SOS programming, such as the application of Positivstellensatz, were proposed in Parrilo (2003). In this way, the convex optimization methods can be used to find appropriate polynomials for the SOS problem. The approximation of nonnegative polynomials by a sequence of SOS was presented in Lasserre (2007). The SOS polynomials incorporate the original nonnegative polynomials in an explicit form.

Prajna et al. (2004) showed sufficient conditions for the solutions to nonlinear control problems, which were formulated in terms of state-dependent Linear Matrix Inequalities (LMI). In the method, the semi-definite programming relaxations based on the SOS decomposition were then used to efficiently solve such inequalities. The application of the SOS decomposition technique to non-polynomial system analysis was summarized in Papachristodoulou and Prajna (2005). Jarvis-Wloszek et al. (2003a) introduced the application of SOS programming to several control problems, e.g., reachable set computation and control design algorithms. A local stability analysis of polynomial systems and an iterative computation method for their region of attraction were presented in Tan and Packard (2008). In Scherer and Hol (2006), the SOS method was applied to two non-convex problems, for example, polynomial semi-definite programming and the fixed-order $\mathscr{H}_2$ synthesis problem.

In Summers et al. (2003), the performance analysis of polynomial systems was presented, by which sufficient conditions were provided for bounds on reachable sets and $\mathscr{L}_2$ gain of nonlinear systems subject to norm-bounded disturbance inputs.

P. Gáspár and B. Németh, *Predictive Cruise Control for Road Vehicles Using Road and Traffic Information*, Advances in Industrial Control,
https://doi.org/10.1007/978-3-030-04116-8

Robust performance in polynomial control systems was analyzed in Topcu and Packard (2009). This method considered the effects of neglected dynamics and parametric uncertainties. Numerical computation problems of convex programming based on the SOS method in practical applications were analyzed in Lofberg (2009). As a new result, the maximum controlled invariant sets of polynomial control systems were calculated in Korda et al. (2013).

The following definitions and theorems are essential to understand SOS programming, see Jarvis-Wloszek et al. (2003a). Let $\mathbb{R}$ denote the real numbers and $\mathbb{Z}_+^n$ denote the set of nonnegative integers. The basic elements of the method are polynomials and SOS as defined below:

Definition B.1 A **Polynomial** f in n variables is a finite linear combination of the functions $m_\alpha(x) := x^\alpha = x_1^{\alpha_1} x_2^{\alpha_2} \cdots x_n^{\alpha_n}$ for $\alpha \in \mathbb{Z}_+^n$, $\deg m_\alpha = \sum_{i=1}^n \alpha_i$:

$$f := \sum_\alpha c_\alpha m_\alpha = \sum_\alpha c_\alpha x^\alpha \tag{B.1}$$

with $c_\alpha \in \mathbb{R}$. Define $\mathscr{R}_n$ to be the set of all polynomials in n variables. The degree of f is defined as $f := \max_\alpha \deg m_\alpha$.

Definition B.2 The **set of SOS** polynomials in n variables is defined as

$$\Sigma_n := \left\{ p \in \mathscr{R}_n \;\middle|\; p = \sum_{i=1}^t f_i^2,\ f_i \in \mathscr{R}_n,\ i = 1, \ldots, t \right\} \tag{B.2}$$

for some $t < \infty$.

A central theorem of SOS programming is Positivstellensatz. By the application of this theorem, the set emptiness constraints of an optimization task can be transformed into SOS feasibility problems.

Theorem B.1 (Positivstellensatz) *Given polynomials* $\{f_1, \ldots, f_r\}$, $\{g_1, \ldots, g_t\}$, *and* $\{h_1, \ldots, h_u\}$ *in* $\mathscr{R}_n$, *the following are equivalent:*

1. *The set*
$$\left\{ x \in \mathbb{R}^n \;\middle|\; \begin{array}{l} f_1(x) \geq 0, \ldots, f_r(x) \geq 0 \\ g_1(x) \neq 0, \ldots, g_t(x) \neq 0 \\ h_1(x) = 0, \ldots, h_u(x) = 0 \end{array} \right\} \tag{B.3}$$
is empty.
2. *There exist polynomials* $f \in \mathscr{P}(f_1, \ldots, f_r)$ *(*$\mathscr{P}$ *is multiplicative convex cone),* $g \in \mathscr{M}(g_1, \ldots, g_t)$ *(*$\mathscr{M}$ *is multiplicative monoid), and* $h \in \mathscr{I}(h_1, \ldots, h_u)$ *(*$\mathscr{I}$ *is ideal) such that*
$$f + g^2 + h = 0. \tag{B.4}$$

The multiplicative monoid and the cone are defined as follows.

Definition B.3 (*Multiplicative monoid*) Given $g_1, \ldots, g_t \in \mathscr{R}_n$, the multiplicative monoid generated by g_j is the set of all finite products of g_j, including the empty product, defined to be 1. It is denoted as $\mathscr{M}(g_1, \ldots, g_t)$. For completeness define $\mathscr{M}(\phi) = 1$.

Definition B.4 (*Cone*) Given $\{f_1, \ldots, f_s\} \in \mathscr{R}$, the cone is generated by f_i is

$$\mathscr{P}(f_1, \ldots, f_s) = \left\{ s_0 + \sum s_i b_i | s_i \in \sum_n, b_i \in \mathscr{M}(f_1, \ldots, f_s) \right\}.$$

For completeness note that $\phi = \sum_n$.

In the practical application of the Positivstellensatz, some assumptions are considered. For example, $g = p_0 \in \mathscr{M}(p_0)$ and $f = -qp_0 - \sum_{i=1}^n s_i p_0 p_i$, $q \in \sum_n$, $p_i \in \mathscr{R}_n$, $h = 0$ are defined. Since q and s_i are SOS, $f \in \mathscr{P}(p_1, \ldots, p_m, -p_0)$. Therefore

$$f + g^2 + h = -qp_0 - \sum_{i=1}^{n} s_i p_0 p_i + p_0^2 = -\left(p_0 - \sum_{i=1}^{m} s_i p_i \right) p_0 - \sum_{i=1}^{n} s_i p_0 p_i + p_0^2 = 0 \tag{B.5}$$

It leads to the generalized S-Procedure, which is formed as follows, see Jarvis-Wloszek (2003).

Theorem B.2 (Generalized S-procedure) *Given symmetric matrices* $\{p_i\}_{i=0}^m \in \mathscr{R}_n$. *If there exist nonnegative scalars* $\{s_i\}_{i=1}^m \in \Sigma_n$ *such that*

$$p_0 - \sum_{i=1}^{m} s_i p_i \succeq q \tag{B.6}$$

with $q \in \Sigma_n$, *then*

$$\bigcap_{i=1}^{m} \{x \in \mathbb{R}^n \mid p_i(x) \geq 0\} \subseteq \{x \in \mathbb{R}^n \mid p_0(x) \geq 0\}. \tag{B.7}$$

The related set emptiness question asks if

$$W := \{x \in \mathbb{R}^n \mid p_1(x) \geq 0, \ldots, p_m(x) \geq 0, -p_0(x) \geq 0, p_0(x) \neq 0\} \tag{B.8}$$

is empty.

Moreover, there is an important connection between SOS programming and LMI problems, which was proved by Parrilo (2003).

Theorem B.3 (LMI feasibility problem) *Given a finite set*

$$\{p_i\}_{i=0}^m \in \mathscr{R}_n,$$

the existence of $\{a_i\}_{i=0}^m \in \mathbb{R}_n$ *such that*

$$p_0 + \sum_{i=1}^m a_i p_i \in \Sigma_n \tag{B.9}$$

is an LMI feasibility problem.

B.2 The Computation Method of Controlled Invariant Sets for Continuous Time Systems

The purpose of the controlled invariant set computation is to determine the impact of limited control actuation on the controllability of the system. Moreover, in several vehicle and traffic systems, there are nonlinearities, which influence the size of the sets. In the following, the goal is to find the largest state-space region, in which the stability of the system can be guaranteed by a given peak-bounded control input. It leads to the computation of controlled invariant sets, see Korda et al. (2013). In this section the problem is examined for continuous time systems, while in the next section, the solution is found for discrete time systems.

The state-space representation of the continuous system is given in the following form:

$$\dot{x} = f(x) + gu, \tag{B.10}$$

where $f(x)$ is a vector which incorporates smooth polynomial functions and $f(0) = 0$. In the next analysis, one control input is considered, thus $u = M_{br}$ or $u = \delta$. The global asymptotical stability of the system at the origin is guaranteed by the existence of the control Lyapunov function of the system defined as follows, see Sontag (1989):

Definition B.5 A smooth, proper and positive-definite function $V : \mathbb{R}^n \to \mathbb{R}$ is a control Lyapunov function for the system if

$$\inf_{u \in \mathbb{R}} \left\{ \frac{\partial V}{\partial x} f(x) + \frac{\partial V}{\partial x} g \cdot u \right\} < 0 \tag{B.11}$$

for each $x \neq 0$.

According to Definition B.6, at the candidate V, two main cases must be distinguished:

1/ If

$$\frac{\partial V}{\partial x} f(x) < 0, \tag{B.12}$$

then the system is stable and $u \equiv 0$. This stability scenario is contained by the next two stability criteria.

2/ If

$$\frac{\partial V}{\partial x} f(x) > 0 \tag{B.13}$$

then the system is unstable. However, the system can be stabilized

2/a: If

$$\frac{\partial V}{\partial x} g < 0 \tag{B.14}$$

and

$$\frac{\partial V}{\partial x} f(x) + \frac{\partial V}{\partial x} g \cdot u_{max} < 0. \tag{B.15}$$

In this case, the upper peak-bound of control input u stabilizes the system.

2/b: If

$$\frac{\partial V}{\partial x} g > 0 \tag{B.16}$$

and

$$\frac{\partial V}{\partial x} f(x) - \frac{\partial V}{\partial x} g \cdot u_{max} < 0. \tag{B.17}$$

In this case, the lower peak-bound of control input u stabilizes the system. Note that $u_{min} = -u_{max}$ is assumed.

The controlled invariant set of the system (B.28) is defined as the level set of the control Lyapunov function at $V(x) = 1$. Thus, the fulfillment of the previous stability criterion must be guaranteed at $V(x) \leq 1$.

Moreover, the Positivstellensatz and Generalized S-Procedure theorems require nonstrict inequality conditions to formulate SOS conditions. Thus, the condition $\frac{\partial V}{\partial x} g < 0$ in 2/a is rewritten to $\frac{\partial V}{\partial x} g \leq -\varepsilon$, where $\varepsilon \in \mathbb{R}^+$ is as small as possible. Similarly in 2/b $\frac{\partial V}{\partial x} g \geq \varepsilon$ is written. Additionally, the conditions $\frac{\partial V}{\partial x} f(x) \pm \frac{\partial V}{\partial x} g \cdot u_{max} < 0$ in 2/a and 2/b are also reformulated to two conditions: $\frac{\partial V}{\partial x} f(x) \pm \frac{\partial V}{\partial x} g \cdot u_{max} \leq 0$ and $\frac{\partial V}{\partial x} f(x) \pm \frac{\partial V}{\partial x} g \cdot u_{max} \neq 0$. Thus, the following inequality conditions are formed:

$$\begin{aligned} \frac{\partial V}{\partial x} g \leq -\varepsilon,\ V(x) \leq 1,\ \frac{\partial V}{\partial x} f(x) + \frac{\partial V}{\partial x} g\ u_{max} \geq 0,\ \frac{\partial V}{\partial x} f(x) + \frac{\partial V}{\partial x} g\ u_{max} \neq 0, \\ \frac{\partial V}{\partial x} g \geq \varepsilon,\ V(x) \leq 1,\ \frac{\partial V}{\partial x} f(x) - \frac{\partial V}{\partial x} g\ u_{max} \geq 0,\ \frac{\partial V}{\partial x} f(x) - \frac{\partial V}{\partial x} g\ u_{max} \neq 0. \end{aligned} \tag{B.18}$$

The conditions (B.18) can be transformed to set emptiness conditions. The advantage of this representation that it can be used in the generalized S-procedure to formulate the SOS existence problem:

$$\begin{aligned}
&\left\{-\frac{\partial V}{\partial x} g \ - \varepsilon \geq 0,\ 1 - V(x) \geq 0,\ L_1(x) \neq 0,\right\} \\
&\left\{\frac{\partial V}{\partial x} f(x) + \frac{\partial V}{\partial x} g\ u_{max} \geq 0,\right\} \\
&\left\{\frac{\partial V}{\partial x} f(x) + \frac{\partial V}{\partial x} g\ u_{max} \neq 0\right\} = \emptyset
\end{aligned} \qquad \text{(B.19)}$$

$$\begin{aligned}
&\left\{\frac{\partial V}{\partial x} g \ - \varepsilon \geq 0,\ 1 - V(x) \geq 0,\ L_2(x) \neq 0,\right\} \\
&\left\{\frac{\partial V}{\partial x} f(x) - \frac{\partial V}{\partial x} g\ u_{max} \geq 0,\right\} \\
&\left\{\frac{\partial V}{\partial x} f(x) - \frac{\partial V}{\partial x} g\ u_{max} \neq 0\right\} = \emptyset
\end{aligned} \qquad \text{(B.20)}$$

Note that the relations in the third inequality are inverted to guarantee the emptiness of the sets. The role of $L_{1,2}(x) \neq 0$ is to guarantee the condition $x \neq 0$ in Definition B.6. $L_{1,2}(x)$ is chosen as a positive-definite polynomial, see Jarvis-Wloszek et al. (2003a). Since it is necessary to find the maximum controlled invariant set, another set emptiness condition is also defined to improve the efficiency of the method. Figure B.1 illustrates that it is predefined a function $p(x)$, of which level set at β value $P_\beta := \{x \in \mathbb{R}^n \mid p(x) \leq \beta\}$ must be inside of the candidate controlled invariant set. Since $p(x)$ is fixed, the maximization of β enlarges P_β together with the controlled invariant set. The condition is defined as

$$\{p(x) \leq \beta,\ V(x) \geq 1,\ V(x) \neq 1\} = \emptyset, \qquad \text{(B.21)}$$

where $p \in \Sigma_n$ is a fixed and positive-definite function.

The previous set emptiness conditions are reformulated to SOS conditions based on the S-procedure. Thus, the next optimization problem is formed to find the maximum controlled invariant set:

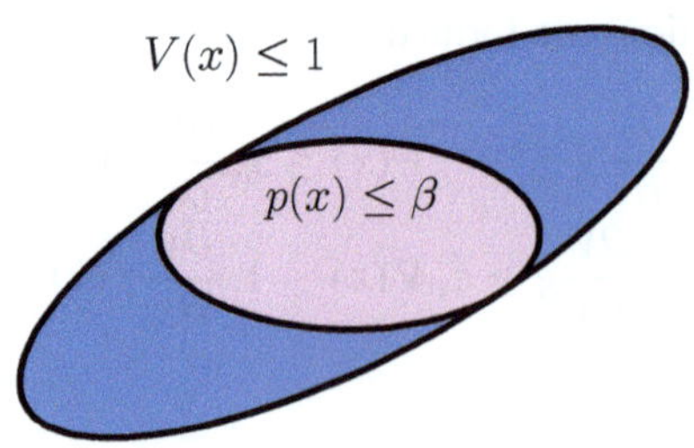

Fig. B.1 Illustration of $p(x)$ and $V(x) = 1$

$$\max \beta \tag{B.22}$$

over SOS polynomials $s_1, s_2, s_3, s_4, s_5 \in \Sigma_n$ and polynomials $V, p_1, p_2 \in \mathscr{R}_n$, $V(0) = 0$
such that

$$\begin{aligned}
&-\left(\frac{\partial V}{\partial x} f(x) + \frac{\partial V}{\partial x} g\, u_{max}\right) - s_1\left(-\frac{\partial V}{\partial x} g\ - \varepsilon\right) - \\
&- s_2(1 - V) - p_1 L_1 \in \Sigma_n && \text{(B.23)}\\
&-\left(\frac{\partial V}{\partial x} f(x) - \frac{\partial V}{\partial x} g\, u_{max}\right) - s_3\left(\frac{\partial V}{\partial x} g\ - \varepsilon\right) - \\
&- s_4(1 - V) - p_2 L_2 \in \Sigma_n && \text{(B.24)}\\
&-(s_5(\beta - p) + (V - 1)) \in \Sigma_n && \text{(B.25)}
\end{aligned}$$

The derivation of (B.23) resembles the one in Jarvis-Wloszek et al. (2003b) but it is more complex because the cone is generated by three terms and there are two polynomials constrained to zero. Some terms of the SOS conditions are omitted in the application of the Positivstellensatz. Although these conditions introduce conservatism, the size of the complexity of the numerical problem is reduced.

B.2.1 Example on the Computation of the Sets

The optimization method of the maximum controlled invariant set has been shown as an illustration of the set computation. This task requires the modeling of the tire forces. Although several tire models have already been published, see, e.g., Pacejka (2004), Kiencke and Nielsen (2000), de Wit et al. (1995), the following polynomial tire modeling approach fits for the set computation problem. In this formalism, the nonlinearities of the tire characteristics can be considered in a given operation range. The nonlinear characteristics of the lateral tire force in the function of tire sideslip α are illustrated in Fig. B.2. The polynomial approximation is formulated as

$$\mathscr{F}(\alpha) = \sum_{k=1}^{n} c_k \alpha^k = c_1 \alpha + c_2 \alpha^2 + \cdots + c_n \alpha^n. \tag{B.26}$$

In the example presented in Fig. B.2, exponent n is chosen 10. Using this approximation, the tire model is valid between $\alpha = -12^\circ \cdots + 12^\circ$.

A polynomial in θ variables of degree $2N$ can be transformed into an LMI with $\binom{\theta+N}{N} \times \binom{\theta+N}{N}$ dimensions, see Parrilo (2003). In the example (Fig. B.2), the degree of the tire model is $2N = 10$, and the system has two variables: α_1 and α_2, thus $\theta = 2$. The size of the LMI is $\binom{2+5}{5}^2 = 42 \cdot 42 = 1764$, which means LMI dimensions. Due to the vast size of the LMI feasibility task, numerical problems may occur. Therefore, the resulting control Lyapunov function V of optimization (B.23) must be checked.

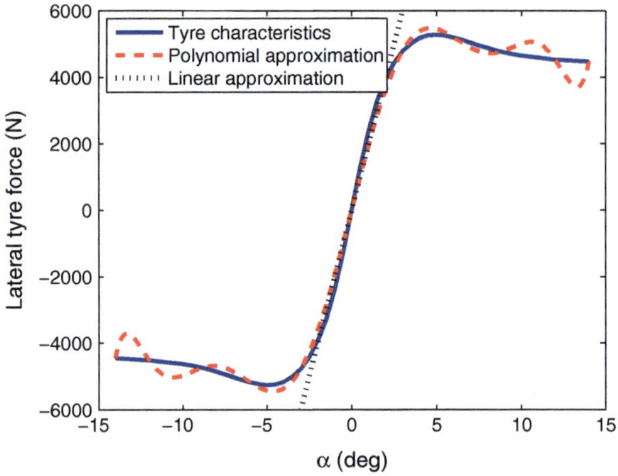

Fig. B.2 The modeling of the lateral tire force

In the following, an alternative computation method is proposed to find the maximum controlled invariant set, which, according to our experience, may lead to an easier calculation. The practical method contains a three-step iterative method.

Step 1: The region of attraction of the uncontrolled system $\dot{x} = f(x)$ is determined as an initial set. In this step, the maximum level set of $V_0 = 1$ is found, which is incorporated in the stable region. The SOS-based computation of region of attraction is presented in Jarvis-Wloszek (2003).

Step 2: An η parameter is chosen and $V_\eta = V_0 \cdot \eta$ is checked as a local control Lyapunov function. The level set $V_\eta = 1$ represents a controlled invariant set S_η, in which the system can be stabilized using a finite control input u. Depending on parameter η, the size of the level set can be enlarged or reduced. The SOS-based computation of local control Lyapunov function is proposed in Tan and Packard (2008). This step is in relation with (B.23), if V is fixed and u is not constrained.

Step 3: In the final step, the acceptability and the enlarging possibility of S_η controlled invariant set must be checked. The peak-bounds of the actuation are $u_{min} = -u_{max}$ and u_{max}. $S_{unst} = \frac{\partial V_0}{\partial x} f(x) > 0$ is the unstable region of the system. $S_{min} = \frac{\partial V_0}{\partial x} f(x) - \frac{\partial V_0}{\partial x} g \cdot u_{max} > 0$ is the region which can not be stabilized by u_{min}. Similarly, $S_{max} = \frac{\partial V_0}{\partial x} f(x) + \frac{\partial V_0}{\partial x} g \cdot u_{max} > 0$ is the region which cannot be stabilized by u_{max}. If S_η is an appropriate controlled invariant set and V_η is an appropriate control Lyapunov function, then

$$S_\eta \bigcap S_{unst} \bigcap S_{min} \bigcap S_{\max} = \emptyset. \tag{B.27}$$

The emptiness of the intersection condition defined above can be checked manually by the plot of S_η, S_{unst}, S_{min}, and $S_{\max}$. Additionally, if S_η is appropriate, then η value can be reduced in the previous step to maximize the controlled invariant set.

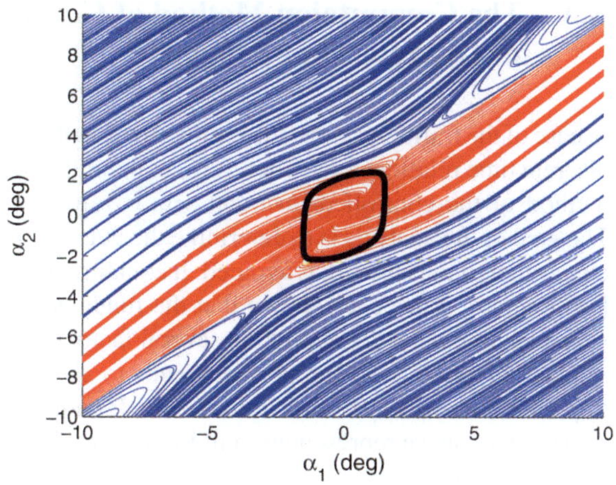

Fig. B.3 Region of attraction set $V_0 = 1$

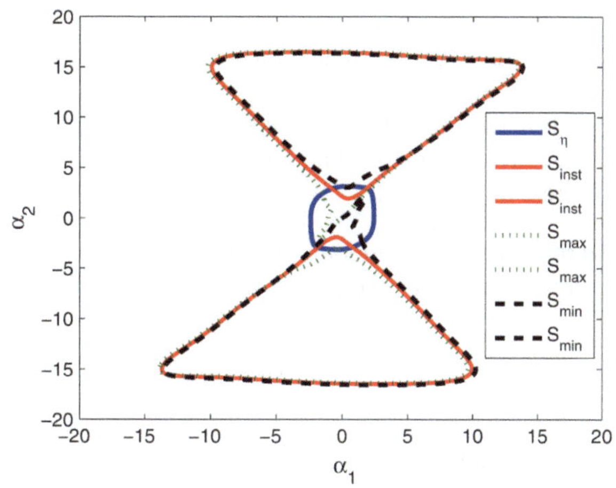

Fig. B.4 The stability regions of the controlled system

Remark B.1 The shape of the maximum controlled invariant set is fundamentally determined by the chosen V_0. If the result of the iterative method V_η is not acceptable, function V_0 of Step 1 should be modified.

The maximum set of the region of attraction is shown in Fig. B.3. In this phase portrait, the red regions are the open-loop stable regions, the blue regions are locally stable $x_\infty \neq \left[0\ 0\right]^T$, and the black set is the region of attraction. This bounding is a conservative approximation, which can be used as an initial set.

In Fig. B.4, the S_η, S_{unst}, S_{min}, and $S_{\max}$ sets of the controlled system are illustrated. The enlargement of S_η is limited in the positive α_2 regions by S_{min}, in the negative α_2 regions by S_{max}.

B.3 The Computaion Method of Controlled Invariant Sets for Discrete Time Systems

In case of the traffic systems, the representation of its dynamics is formulated through discrete time relationships. Therefore, in the following, the SOS-based computation method of the controlled invariant sets is reformulated to discrete time systems. The goal of the analysis is to determine the maximum inflow $q_{in,max}(k)$, at which the congestion of the network is avoided. It contains all of the inflow in the traffic system. Furthermore, it has been shown that the fundamental diagram of the traffic system varies significantly, depending on several factors. This variation is represented by a scheduling variable ρ. The problem leads to an optimization process, in which the $u_{max}(\rho(k)) = q_{in,max}(k)$ function must be found.

The state-space representation of the system (7.6) with $u_{max}(\rho(k))$ is given in the following form:

$$x(k+1) = f(\rho(k), x(k)) + g u_{max}(\rho(k)), \tag{B.28}$$

where $f(\rho, x(k))$ is a matrix, which incorporates smooth polynomial functions and $f(\rho, 0) = 0$.

The local stability of the system at the origin is guaranteed by the existence of the control Lyapunov function of the system, see Sontag (1989). It is rewritten to discrete time case as

Definition B.6 A smooth, proper, and positive-definite function $\mathrm{V} : \mathbb{R}^n \to \mathbb{R}$ is a control Lyapunov function for the system (B.28) if

$$\Delta \mathrm{V}(\rho(k), x(k))\Big|_{u(k)=u_{max}(\rho(k))} < 0 \tag{B.29}$$

for each $x(k) \neq 0$ and $u_{max}(\rho(k)) \geq 0$, $\mathrm{V}(\rho(k), 0) = 0$ for all ρ.

The parameter-dependent control Lyapunov function is chosen in the next form:

$$\mathrm{V}(\rho(k), x(k)) = V(x(k)) \cdot b(\rho(k)) \tag{B.30}$$

where $b(\rho(k))$ is an intuitively chosen ρ dependent basis function. The difference in the condition (B.29) is expanded using (B.30)

$$\begin{aligned} \Delta \mathrm{V}(\rho(k), x(k)) &= \mathrm{V}(x(k+1), \rho(k)) - \mathrm{V}(x(k), \rho(k)) + \\ &+ \mathrm{V}(x(k), \rho(k+1)) - \mathrm{V}(x(k), \rho(k)) = \\ &= (V(x(k+1)) - V(x(k))) \cdot b(\rho(k)) + \\ &+ (b(\rho(k+1)) - b(\rho(k))) \cdot V(x(k)) = \\ &= \Delta V(x(k)) \cdot b(\rho(k)) + \Delta b(\rho(k)) \cdot V(x(k)) \end{aligned} \tag{B.31}$$

During the decomposition of $x(k)$ and $\rho(k)$ in the control Lyapunov function, the original difference $\Delta \mathrm{V}(\rho(k), x(k))$ is separated to $\Delta V(x(k))$ and $\Delta b(\rho(k))$. The difference $\Delta V(x(k))$ using (B.28) is computed as

$$\Delta V(x(k)) = V(f(\rho(k), x(k)) + gu(k)) - V(x(k)) \tag{B.32}$$

Considering the function $u_{max}(\rho(k))$ in (B.32):

$$\begin{aligned} \Delta V(x(k))\Big|_{u(k)=u_{max}(\rho(k))} &= \\ &= V(f(\rho(k), x(k)) + gu_{max}(\rho(k))) - V(x(k)) \end{aligned} \tag{B.33}$$

The difference $\Delta b(\rho(k))$ depends on the intuitively formulated basis function $b(\rho(k))$. In the further computation, the upper limit of $\Delta b(\rho(k))$ is used, such as $\nu \geq |\Delta b(\rho(k))|$, see Wu (1995). The upper limit is chosen according to the dynamic features of the traffic system.

The local stability criterion (B.29) determines the set of the states x, which are stable and yields that $x_\infty < \infty$, if u_{max} is applied. This set is bounded by the controlled invariant set, defined in the following way:

$$\mathrm{V}(\rho(k), x(k)) = 1 \tag{B.34}$$

For practical reasons, the equality definition (B.34) is substituted for by two inequality conditions:

$$1 - \varepsilon \leq \mathrm{V}(\rho(k), x(k)) \leq 1, \tag{B.35}$$

where $\varepsilon > 0$ is an infinitesimally small number.

Further constraints on the stabilization (B.29) are the validity ranges of the scheduling variable ρ and the state x. The choice of the basis function $b(\rho(k))$ is valid in a range

$$\rho_{min} \leq \rho(k) \leq \rho_{max}, \tag{B.36}$$

where ρ_{min} and ρ_{max} are the bounds of the scheduling variable. Moreover, the solution of the urban network gating must be found at the constraint:

$$0 \leq x(k) \tag{B.37}$$

which represents that the number of vehicles in the network is positive or zero.

The local stability condition (B.29) is constrained by the controlled invariant set (B.35), the constraints of the scheduling variable (B.36) and the states (B.37). The stability problem with the constraints is transformed into a set emptiness conditions:

$$
\Big\{(V(f(\rho(k),x(k))+gu_{max}(\rho(k)))-V(x(k)))\cdot b(\rho(k))+ \\
+\nu\cdot V(x(k))\geq 0, \\
(V(f(\rho(k),x(k))+gu_{max}(\rho(k)))-V(x(k)))\cdot b(\rho(k))+ \\
+\nu\cdot V(x(k))\neq 0, \\
V(x(k))\cdot b(\rho(k))-(1-\varepsilon)\geq 0, \\
1-V(x(k))\cdot b(\rho(k))\geq 0, \quad x(k)\geq 0 \\
\rho(k)-\rho_{min}\geq 0, \quad \rho_{max}-\rho(k)\geq 0,\Big\}=\emptyset \tag{B.38}
$$

Using the generalized S-Procedure, the set emptiness condition (B.38) is transformed into an SOS condition. The optimization problem is to find an $u_{max}(\rho(k))$ solution and feasible $\mathrm{V}(\rho(k),x(k))$ for the following task:

$$
\max\ u_{max}(\rho(k)) \tag{B.39}
$$

over $s_1,s_2,s_3,s_4,s_5\in\Sigma_n;\ V(x(k)),b(\rho(k))\in\mathscr{R}_n$
such that

$$
-\bigg((V(f(\rho(k),x(k))+gu_{max}(\rho(k)))- \\
-V(x(k)))\cdot b(\rho(k))+\nu\cdot V(x(k))\bigg)- \\
-s_1\bigg(V(x(k))\cdot b(\rho(k))-(1-\varepsilon)\bigg)- \\
-s_2\bigg(1-V(x(k))\cdot b(\rho(k))\bigg)-s_3x(k)- \\
-s_4\left(\rho(k)-\rho_{min}\right)-s_5\left(\rho_{max}-\rho(k)\right)\in\Sigma_n. \tag{B.40}
$$

B.4 LPV-Based Computation Method of Controllability Sets for Continuous Time Systems

Finally, an alternative solution for the computation of the controlled invariant sets is presented. In this process, the controllability set computation is based on the trajectory reversing method, see Horiuchi (2015). It means that the null controllability region of the forward time nonlinear system is equivalent to the reachability region of the reverse time system, see Snow (1967). Thus, an alternative solution can be the determination of the reachability sets of the reverse time system. In the following, the reachability set computation for LPV systems is presented.

The continuous time LPV system is formed as

$$\dot{x} = A(\rho)x + Bu, \tag{B.41}$$

where ρ is the scheduling variable, x is the state of the system, and u is the control input.

The reachable sets of the vehicle systems are approximated by using ellipsoidal forms. According to preliminary analysis, the ellipsoid form is found as an appropriate selection. Boyd et al. (1997) presents the conditions to find the minimal reachable set for LDIs and linear systems. The LMI condition of the reachable set is formulated as

$$\begin{bmatrix} A^T P(\rho) + P(\rho)A + \alpha P(\rho) + \dot{P}(\rho) & P(\rho)B_2 \\ B_2^T P(\rho) & -\alpha I \end{bmatrix} \leq 0, \tag{B.42}$$

if there exists $P(\rho)$ and α satisfying $P(\rho) > 0, \alpha \geq 0$. The Lyapunov function of the system is chosen as in a parameter-dependent way $V(x, \rho) = x^T P(\rho)x$, and time derivative of $P(\rho)$ is handled as $\dot{P}(\rho) = \frac{\partial P(\rho)}{\partial \rho}\dot{\rho}$. The upper limit of $\dot{\rho}$ is predefined as $\nu \geqq |\dot{\rho}|$ and $\frac{\partial P(\rho)}{\partial \rho}$ is computed by using the formulation

$$P(\rho) = \sum_{i=1}^{N} f_i(\rho)P_i \qquad P_i \in \mathscr{R}^{nxn} \tag{B.43}$$

and f_i are appropriately chosen basis functions. In the solution of the LMI feasibility problem, it is necessary to find an α value in which $log(det(P(\rho)^{-1}))$ is minimal and which represents the volume over an ellipsoidal cylinder. In the proposed method, the solution of (B.42) is the ellipsoidal approximation of the reachable set: $\varepsilon = \{x|x^T P(\rho)x \leq 1\}$.

According to our experience, the computation of $P(\rho)$ may be numerically difficult because of the LMI condition (B.42). Moreover, the selection of f_i basis functions in (B.43) determines the qualification of reachable set approximation. However, the formulation of the basis function is not a trivial task. Therefore, in the following, a computation method for an outer approximation of the reachable set is proposed. Simulation experiments are used in order to formulate the shape of the reachable set. The evaluation of the shape provides preliminary information in the approximation of the reachable set. Thus, the simulations can help to determine the structure of $P(\rho)$, see, e.g., Németh and Gáspár (2013a).

Based on the simulation experiments, the selected $P \leftharpoonup \rho \leftharpoondown \psi$ must be verified whether it is the solution of LMI (B.42). The advantage of this method is that it is not necessary to solve the original LMI feasibility problem, because $P(\rho)$ is already known. If $P(\rho)$ represents the reachable set, then all of the eigenvalues of the following matrix are negative:

$$max\left(eig\left(\begin{bmatrix} A^T P + PA + \alpha P + \dot{P} & PB_2 \\ B_2^T P & -\alpha I \end{bmatrix}\right)\right) < 0 \tag{B.44}$$

Note that (B.44) contains an $\alpha \geq 0$ parameter, which is not known in the computation of eigenvalues. In addition, $P(\rho)$ function contains some parameters ν_i, which are also not known. In the calculation of the reachable set it is necessary to find the smallest outer approximation. Thus, $P(\rho) = P(\rho, \nu_i)$ function must be optimized to provide the upper bound of the reachable set.

The optimization task is to find the solution

$$\min_{\nu_i, \alpha} \; log \left(det \left(P(\rho, \nu_i)^{-1} \right) \right), \tag{B.45}$$

while the constraints (B.44) and $\alpha \geq 0$ are fulfilled. Using the proposed method, the numerically complex LMI feasibility problem, see Boyd et al. (1997); Shin (2002) is replaced by a constrained optimization problem. There are several methods to solve the constrained optimization, see Gill et al. (1981), Byrd et al. (2000).

The numerical solution is based on the Nelder–Mead simplex direct search algorithm, see Lagarias et al. (1998). It uses $n + 1$ points for n-dimensional vectors X. At the beginning of the optimization, the algorithm makes a simplex around an initial guess X_0 by adding a small value of each component $X_0(i)$ and using these vectors as elements of the simplex. During the optimization, the algorithm modifies the simplex repeatedly according to the optimization. The advantages of the replacement of LMI solution are that the original numerically difficult computation procedure is replaced by an easier solution and the complex nonlinear basis functions are substituted for using simulation experiments.

Remark B.2 The simulation experiments have an important role in the approximation of the reachable set. However, it is necessary to emphasize that the proposed method always results in an outer approximation of the reachable set. The importance of the appropriate simulation is the evaluation of the shape of the reachable set. It helps formulate a suitable formulation of reachable set approximation.

The initial step of reachable set approximation is to create appropriate simulation examinations. For example, in case of lateral vehicle dynamic simulations, both the steering and the brake systems are analyzed and the input signals are selected in their entire amplitude and frequency range. Two kinds of signals are used as excitations, i.e., chirp (sinusoidal with linearly increasing frequency) and pulse signals. During simulations, the velocity of the vehicle must also be changed because the nonlinear effect of velocity is analyzed.

The formulation of the parameter-dependent Lyapunov function $P(\rho)$ is the following. The volume and rotation of elliptical cylinder change to the effect of velocity, therefore this information is used in the formulation of $P(\rho)$. The description of $P(\rho)$ is used in the approximation of the reachable set:

$$P(\rho) = T^T(\rho) P_0 T(\rho), \tag{B.46}$$

where P_0 is a level of the elliptical cylinder. $T(\rho)$ is a transformation matrix, which contains both volume and rotation:

$$T(\rho) = \left(1 + \nu_1 \left(\frac{\rho}{\rho_0} - 1\right)\right) \begin{bmatrix} \cos(\nu_2 \rho) & -\sin(\nu_2 \rho) \\ \sin(\nu_2 \rho) & \cos(\nu_2 \rho) \end{bmatrix}, \tag{B.47}$$

where ν_1 and ν_2 are parameters and ρ_0 is the value of the scheduling variable at P_0 level.

References

Apkarian P, Gahinet P (1995) A convex characterization of gain-scheduled $\mathscr{H}_\infty$ controllers. IEEE Trans Autom Control 40(5):853–864

Becker G, Packard A, Philbrick D, Balas G (1993) Control of parametrically dependent linear systems: a single quadratic Lyapunov approach. In: American control conference, San Francisco, pp 2795–2799

Bokor J, Balas G (2005) Linear parameter varying systems: a geometric theory and applications. In: 16th IFAC world congress, Prague 1(1–12)

Boyd S, Ghaoui LE, Feron E, Balakrishnan V (1997) Linear matrix inequalities in system and control theory. Society for industrial and applied mathematics, Philadelphia

Byrd R, Gilbert JC, Nocedal J (2000) A trust region method based on interior point techniques for nonlinear programming. Math Program 89(1):149–185

de Wit CC, Olsson H, Astrom KJ, Lischinsky P (1995) A new model for control of systems with friction. IEEE Trans Autom Control 40(3):419–425

Gahinet P, Apkarian P (1994) A linear matrix inequality approach to $\mathscr{H}_\infty$ control. Int J Robust Nonlinear Control 4:421–448

Gahinet P, Apkarian P (1996) Explicit controller formulas for LMI-based $\mathscr{H}_\infty$ synthesis. Automatica 32:1007–1014

Gill PE, Murray W, Wright M (1981) Practical optimization. Academic Press, London

Horiuchi S (2015) Evaluation of chassis control algorithms using controllability region analysis. In: International symposium on dynamics of road vehicles on roads and tracks, Graz, Austria

Jarvis-Wloszek Z (2003) Lyapunov based analysis and controller synthesis for polynomial systems using sum-of-squares optimization. PhD Thesis, University of California, Berkeley

Jarvis-Wloszek Z, Feeley R, Tan W, Sun K, Packard A (2003a) Some controls applications of sum of squares programming. In: 42nd IEEE Conference on Decision and Control, vol 5. Maui, USA, pp 4676–4681

Jarvis-Wloszek Z, Feeley R, Tan W, Sun K, Packard A (2003b) Some controls applications of sum of squares programming. In: 42nd IEEE international conference on decision and control (IEEE Cat. No.03CH37475), vol 5, pp 4676–4681

Kiencke U, Nielsen L (2000) Automotive control systems for engine, driveline and vehicle. Springer

Korda M, Henrion D, Jones CN (2013) Convex computation of the maximum controlled invariant set for polynomial control systems. In: Conference on decision and control, pp 7107–7112

Lagarias J, Reeds JA, Wright MH, Wright PE (1998) Convergence properties of the nelder-mead simplex method in low dimensions. SIAM J Optim 9(1):112–147

Lasserre JB (2007) Sum of squares approximation of nonnegative polynomials. SIAM J Optim 49(4):651–669

P. Gáspár and B. Németh, *Predictive Cruise Control for Road Vehicles Using Road and Traffic Information*, Advances in Industrial Control,
https://doi.org/10.1007/978-3-030-04116-8

Lofberg J (2009) Pre- and post-processing sum-of-squares programs in practice. IEEE Trans Autom Control 54(5):1007–1011

Németh, Gáspár P (2013) Analysis of vehicle actuators based on reachable sets. In: 2013 European control conference (ECC), pp 3137–3142

Nemirovsky A, Gahinet P (1994) The projection method for solving linear matrix inequalities. In: Proceedings of the American control conference

Nesterov Y, Nemirovsky A (1993) Interior point polynomial methods in convex programming: theory and applications. SIAM

Pacejka HB (2004) Tyre and vehicle dynamics. Elsevier Butterworth-Heinemann, Oxford

Packard A, Becker G (1992) Quadratic stabilization of parametrically dependent linear systems using parametrically dependent linear feedback. Adv Robust Nonlinear Control Syst 43:29–36

Packard A, Balas G (1997) Theory and application of linear parameter varying control techniques. In: American control conference, Workshop I, Albuquerque, New Mexico

Papachristodoulou A, Prajna S (2005) Analysis of non-polynomial systems using the sum of squares decomposition. In: Positive polynomials in control. Springer

Parrilo P (2003) Semidefinite programming relaxations for semialgebraic problems. Math Program Ser B 96(2):293–320

Prajna S, Papachristodoulou A, Wu F (2004) Nonlinear control synthesis by sum of squares optimization: a lyapunov-based approach. In: Proceedings of the 5th IEEE Asian control conference, vol 1, pp 157–165

Rough W, Shamma J (2000) Research on gain scheduling. Automatica 36:1401–1425

Scherer CW, Hol CWJ (2006) Matrix sum-of-squares relaxations for robust semi-definite programs. Math Program 107:189–211

Scherer C, Gahinet P, Chilali M (1997) Multiobjective output-feedback control via LMI optimization. IEEE Trans Autom Control 42(7):896–911

Shamma J, Athans M (1991) Guaranteed properties of gain sheduled control of linear parameter-varying plants. Automatica 27:559–564

Shin J (2002) Analysis of linear parameter varying system models based on reachable sets. In: American control conference, Anchorage, vol 1, pp 35–40

Snow DR (1967) Determining reachable regions and optimal controls. In: Advances in Control Systems, vol 5. Academic Press

Sontag ED (1989) A "universal" construction of Artstein's theorem on nonlinear stabilization. Syst Control Lett 13:117–123

Summers E, Chakraborty A, Tan W, Topcu U, Seiler P, Balas G, Packard A (2003) Quantitative local l_2-gain and reachability analysis for nonlinear systems. Int J Robust Nonlinear Control 23(10):1115–1135

Tan W, Packard A (2008) Stability region analysis using polynomial and composite polynomial lyapunov functions and sum-of-squares programming. IEEE Trans Autom Control 53(2):565–571

Topcu U, Packard A (2009) Local robust performance analysis for nonlinear dynamical systems. In: Proceedings of the American control conference, pp 784–789

Wu F (1995) Control of linear parameter varying systems. PhD Thesis, Mechanical Engineering, University of California at Berkeley

Wu F (2001) A generalized lpv system analysis and control synthesis framwework. Int J Control 74:745–759

Wu F, Yang X, Packard A, Becker G (1996) Induced L_2 norm controller for LPV systems with bounded parameter variation rates. J Robust Nonlinear Control 6:983–988

Index

P. Gáspár and B. Németh, *Predictive Cruise Control for Road Vehicles Using Road and Traffic Information*, Advances in Industrial Control,
https://doi.org/10.1007/978-3-030-04116-8

MIX
Papier aus verantwortungsvollen Quellen
Paper from responsible sources
FSC® C105338

If you have any concerns about our products,
you can contact us on
ProductSafety@springernature.com

In case Publisher is established outside the EU,
the EU authorized representative is:
Springer Nature Customer Service Center GmbH
Europaplatz 3, 69115 Heidelberg, Germany

Printed by Libri Plureos GmbH
in Hamburg, Germany